BRIDGE
INNOVATION

BRIDGE
INNOVATION
Technique that tranformed past failure into success.

BHOLA SHANKAR PANDEY

PARTRIDGE
A Penguin Random House Company

Print information available on the last page.

To order additional copies of this book, contact
Partridge India
000 800 10062 62
orders.india@partridgepublishing.com

www.partridgepublishing.com/india

AUTHOR NOTE

I take immense pleasure and pride to dedicate this book **"Bridge Innovation - Technique that transformed past failure into success"** to my beloved Nation.

The idea to write this book was first suggested by Shri SA Reddi. Further, I took opportunity to build and structure my thoughts to work and complete a draft version of this book. Shri S A Reddi reviewed the draft book and provided valuable suggestions. In addition, Shri DD Sharma - chairman of D2S infrastructure, also reviewed the draft and provided his valuable input. I express my gratitude to both of them.

After 1st Draft, the book was reviewed and proof edited by Er Vinit Vishal - B. Tech (EE), MS (IT). Further, Vinit Vishal, with admirable meticulousness, worked closely with publisher and went through several iteration on content and design to reshape and deliver this book. I thank him for all his dedicated hard work and sincere efforts.

Smt Ragini Singh and Er Vandana Kumari has also put their best to produce the index to facilitate an early publication. I thank them as well. I am also deeply indebted to many others who helped in many other ways to bring out this book.

I will be grateful to the readers for their suggestions in improving this further and look forward to their valuable feedback.

B. S. PANDEY
B. Sc Engg (Civil),
M. E (Civil), CE (IND)
MIE (IND) LMIRC)

PREFACE OF THE EDITION

I thank to almighty who provided me strength and vision to compile and complete this book. This book has been compiled primarily for the "Practical civil workers" and should prove useful to civil engineers, Architects, Consultants of the various agencies. The objective of this book is to give a complete and concise account of the various solutions to serve as a ready reference for problems constantly confronted by civil engineers.

We have made all possible efforts to make this book comprehensive, interesting and complete. Our focus has always been to elucidate complex engineering principles in simple and plain practical language.

My sincere thanks to my numerous colleagues and friends for the valuable help given me in my work. I appreciate Shri Din Dayal Sharma - Chairman D2S Infrastructure Pvt Ltd and Shri S A Reddi - Chief consultant, for the valuable help in data collection for the modern methods of construction, review and correction on the contains of this book.

I express my special gratitude and thanks to Tapeshwari Pandey (late father), Sampati Devi (late mother), Smt Ragini Singh (Wife), Er Vinit Vishal (Son), Dr Gaurav Vishal (Son), Dr Abhishek Ranjan (son in law), Dr Vanita Vaishali (Daughter), Er Vandana Kumari & Dr Sweta Shekhar (daughter in law) for aspiration, support and help. Special thanks to my lovely grandsons Vilohit Pandey, Master Aarav and Granddaughter Ira, whose presence around bring happiness and fill me with energy.

I will be grateful and appreciate those who will kindly call attention to any errors or omission or commission or give valuable suggestions for improvement of the book to enhance its usefulness.

CONTENTS

ACKNOWLEDGEMENTS

The authors express their gratitude to:

1. Shri R. B. Singh, the designer, and Mr Rajeev Ahuja of Arch Infrastructures Pvt Ltd, the proof consultant, for the feedback and guidance on the relevant IRC and ISO codes. They have used past experience in the field of civil engineering since May 1981.
2. The grace of God Almighty.
3. Lt. Gen. A. K. Nanda, then DGBR, for his dynamic leadership during his present tenure while completing the bridge.
4. Lt. Gen. K. S. Rao, DGBR (retd), who took a bold and dynamic decision to (*a*) abandon the past structures and (*b*) accept a new very long span of 160 m long.
5. Dr V. K. Yadav, then DDG/bridging, and his brilliant team of engineering officers, who ensured no hindrances in the work and assured approval of design and drawings. The team was involved from the inception of the project scheme till its completion.
6. CE (P) Sampark and his earlier and present team along with Commander, 13 BRTF and all members of BRTF and the RCCs of 57, 69, 104, and allied units.
7. S. K. Chellani, project management consultants, and his team for the entire execution of the work.
8. M/s Usha Martin Ltd (2A Shakespeare Sarani, Kolkata 700 071) for the timely supply of LRPC HT strands.
9. M/s Vikrant Ispat Udyog (113/8 Navyug Market, First Floor, Ghaziabad 201001) for giving all steel, be it for reinforcement or structural steel, on time.
10. All members of the project team of D2S Infrastructures Pvt Ltd, led by Shri Y. Singh, the project manager.
11. M/s Jaiprakash Associates Ltd, Cement Division (64/4 Site, No. IV, Sahibabad Industrial Area, Ghaziabad 201010, UP) for giving good-quality OPC cement always on time; thereby, there was no loss of time during the execution.

12. Last but not the least, Shri Krishna Biswas, the labour contractor, for his technical brilliance and for his untiring efforts in achieving the six-day cycle through all types of weather, be it extreme cold, rain, or sunshine.

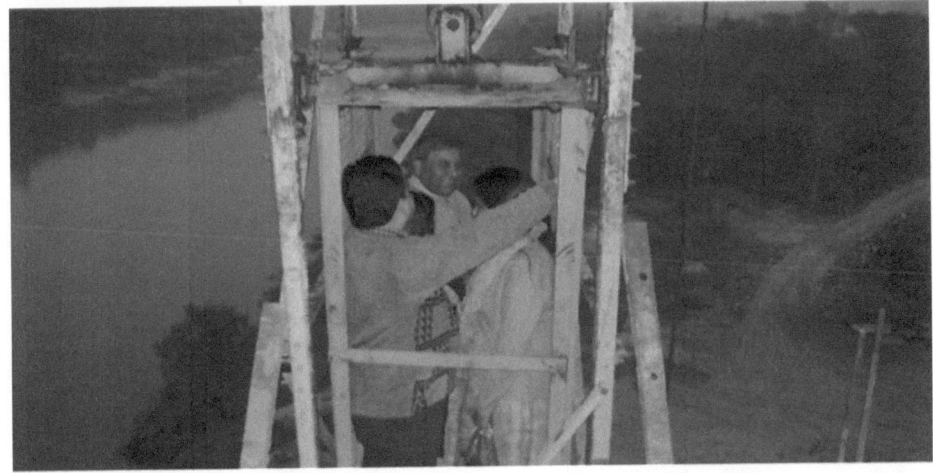

1.1 INTRODUCTION

The construction of bridges over any river is a most difficult and challenging job requiring super-skilled civil engineers since the creation of earth by nature. Our religious book *Ramayana* also describes the difficulties on the construction of a bridge while Purushottam Ramachandran decided to fight with Ravana, king of Sri Lanka, during Satyug; the great difficulties for Ramachandran forced him to cross the sea. After failure of all means, God called expert engineers Shri Nal and Niel for the construction of a bridge over the sea known as Rameshwaram Bridge.

With the inspiration of God and *dibya driti* of Mata Vaishno Devi, a new concept for a bridge scheme was imagined and proposed during August 2005 to the director general of Border Roads Organisation (BRO) after 26 years of continuous failure in the construction of the Chenab Bridge by the author at the same site, where the bridge was partly constructed but failed near the town of Akhnoor. The detailed analysis for the reasons of failure was ascertained, and several historical faiths were studied as well before finally a proposal for a bridge scheme which suits the site was projected.

The proposal was analyzed by several experts, and finally, the bridge scheme was approved for construction through outsourcing for a total length of 280 m with the largest single span of 160 m to avoid construction of any structure in the main stream of the river due its hostile and turbulent current and conglomerate strata. This is the first and largest single-span, prestressed, continuous cantilever bridge designed and constructed in India so far.

1.2 THE RIVER CHENAB

The river Chenab is one of the biggest rivers in India. This river has been regarded with religious sentiment. The origin of this river is in Tandi in the upper Himalayas, at the Lahaul and Spiti District of Himachal Pradesh; it is the merging place of two streams—the Chandra and Bhaga—and forms the river Chenab. The birth of the river Chenab has other religious stories also. Chandra and Bhaga are named after two rishis who had gone to the peak of the upper Himalayas on the opposite side of Baralacha Pass at the altitude of 16,405 ft and started sadhana as well as dhyana to the God Shiva. Due to sadhana, both rishis had come to know that there is need for sufficient water in this area for habitants. Accordingly, the rishis generated two streams of water from Chandra and Bhaga; these met at Tandi to form a river called Chandrabhaga. Therefore, in the upper plain of Himachal Pradesh, it is also known as Chandrabhaga. It flows through the Jammu region of Jammu and Kashmir (J & K) and into the plain of Punjab. The river Chenab literally means 'moon [*chen*]' and 'river (*aab*)' and flows from the upper Himalayas in the Lahaul and Spiti District of Himachal Pradesh, India.

Fig. 1: General view of the bridge site before construction.

The river Chenab in Punjab constitutes the bounds between the Rechna and the Jech interfluves. The Ravi and the Jhelum rivers join the Chenab at Trimmu. Near Uch Sharif, it merges with the Sutlej River to form the famous five rivers of Punjab. The Beas River joins the Sutlej River near Ferozepur, India. The Sutlej joins the Indus at Mithankot. The river Chenab is approximately 960 km in length. Total length of the river is approximately 1,200 km, capturing a catchment area of 61,000 km² in Himachal, the foothills of J & K and the Punjab provinces of Pakistan before joining Sutlej. The waters of the Chenab are shared by India and Pakistan as per the terms of the Indus Water Treaty. It flows through the Jammu region of Jammu and Kashmir into the plains of Punjab.

Chenab is known by Indians as Ashkini or Iskmati in the Vedic period and as Acesines by the ancient Greeks. In 325 BC, Alexander the Great supposedly founded the town of Alexandria on the Indus, which is the present-day Uch Sharif, Mithankot, or Chacharan at the merging of the Indus and the united stream of Punjab rivers, which is currently known as the Panjnad River.

The Chenab is perceived by the people of the Punjab in the same way as, say, the Rhine for the Germans or the Danube for the Austrians and the Hungarians. It is the vital river around which Punjabi rituals revolve, and it plays an important part in the story of *Heer Ranjha*, the Punjabi national epic, and *Sohni Mahiwal*.

The river has been in the news of late due to the steps taken by the Indian government to build a number of hydropower dams along its length in India, and the most notable is the Baglihar hydropower project.

The river Chenab has a heavy flow of current. The water is extremely chilly, and the river is very wide in width and is deep in the Jammu and Kashmir area. In 1915 a historical book has described that the river Chenab is not suitable for the construction of any structure in its main stream due the high velocity of the water and the conglomerate strata apart from the deep water depth as well as the very chilly water.

1.3 THE HISTORICAL TOWN OF AKHNOOR

Akhnoor is a town in Jammu district in the state of Jammu and Kashmir, India, situated on the bank of the river Chenab and very near to the Pakistan towns of Chamm and Joria. It is a picturesque town approximately 32 km away from Jammu to the north-west of the city. The town is located at the coordinates of 32.87° N 74.73° E 32.87; 74.73 (S), having a time zone of IST (4TC + 5.30), an elevation of 301 m (988 ft), pin code of 181 201, and telephone code of +911924.

This beautiful town has historical importance; there are some ruins from the Indus Valley civilization in the town of Akhnoor along the river Chenab, such as a cave where the Pandavas were believed to have hidden. The popular story of how the town Virat Nagar got its name converted to Akhnoor is that she was prescribed to wash her eyes with the holy water of the river Chenab using some Ayurvedic medicine from a local Hindu priest. Her vision was regained and hence the name. In Urdu, the word *noor* means 'vision/glow/shine', and the word *aankh* means 'eye'. So her eyes regained their vision and hence the name Akhnoor.

The people of this town had faced a number of problems since civilization. The earlier problem was crossing the Chenab for connecting Jammu and the rest of India; it has been partly resolved by the construction of the 208 m three-span (1 × 138 m + 2 × 35 m) steel bridge by Maharaja Hari Singh in 1932, which was named Yuvaraj Karan Singh Bridge. The bridge had helped the people access Jammu from Jouria, Palanwalla, Akhnoor, etc. by surface communication.

The people of this town had fled during Indo-Pakistani War in 1965 and 1971 and returned back gradually after the ceasefire. The Pakistan border is around 18 km away by road; aerial distance is only 8 km. Further, people had suffered badly in 1992 when the main span of the steel bridge over the river Chenab was washed away on 10 September 1992. The bridge was rebuilt and opened for traffic on 13 April 1994 by the Border Roads Organisation for Class 18R loading.

Akhnoor town has a population of (approximately) 25,000. Males constitute 55% of the population, and females constitute 45%. The literacy of this town is 80% higher than the national average of 59.50%, with approximately 55% males and 45% females. Approximately 15% of the population is under 6 years of age. The languages spoken are Dogri, followed by Punjabi, Urdu, Hindi, and English.

1.4 HISTORY OF JAMMU, THE NEAREST CITY

Jammu is separated from Kashmir by the Pir Panjal mountain range. Many historians and locals believe that Jammu was founded by Raja Jambu Lochan in fourteenth century BC. During one of his hunting campaigns, he reached the Tawi River, where he saw a goat and a lion drinking water at the same place. Having satisfied their thirst, they went their own ways. The raja was amazed, and he abandoned the idea of hunting and returned to his companions. Expressing what he had seen, he exclaimed that this place was a place of peace and tranquillity, where a lion and a lamb could drink water side by side. The raja commanded a palace to be built at this place and a city to be founded around it.

Thus, a city was developed and was named Jambu Nagar, which then later changed into Jammu. Jambu Lochan was the brother of Raja Bahu Lochan, who had constructed a fort on the bank of the river Tawi. Bahu Fort is one of the famous historical places in Jammu.

The city is mentioned in the ancient book Mahabharata. Excavations near Akhnoor, 20 mi (32 km) from Jammu City, provide evidence that Jammu was once a part of the Harappan civilization.

Remains from the Maurya, Kushan, Kushanshahs, and Gupta periods have also been found in Jammu. After AD 480 the area was dominated by the Hephthalites, who ruled from Kapisa/Kabul. They were succeeded by the Kushano–Hephthalite dynasty from AD 565 to 670, then by the Shahis from 670 to the early 1000s, when the Shahis were destroyed by the Ghaznavids.

Jammu is also mentioned in the campaigns of Timur. The area witnessed the change of control from the invading Mughals and Sikhs before finally falling under the control of the British. After independence, it became a part of the Republic of India, the direct successor of India itself, following a bitter Kashmir war.

The Bhalessa tract enveloping the adjoining hills of Chamba in Himachal Pradesh is inhabited by the Gaddis, another semi-nomadic community who graze immense flocks of goats and sheep along the Himalayan slopes. As summer draws on, the Gaddis move up the mountain pastures with their flocks and return to the lower area with the first snowfall. Gaddis are generally associated with emotive music played on the flute.

Jammu, also known as the City of Temples, has innumerable temples and shrines with glittering shikharas soaring into the sky; they dot the city's skyline, creating the ambiance of a holy and peaceful city.

Home to some of the most popular Hindu shrines, such as Vaishno Devi, Jammu is one of the most famous tourist destinations for pilgrimages in India. Once the seat of power for the Dogra Rajput dynasty, Jammu came under the control of Maharaja Ranjit Singhji in the nineteenth century and became a part of the Sikh Empire. Maharaja Ranjit Singh soon appointed Gulab Singhji as the ruler of Jammu. After the death of Maharaja Ranjit Singh, Punjab, the Sikh Empire, was defeated by the British after Maharaja Duleep Singh was taken by the British to England under the orders of the company. Not having the resources to occupy the hills immediately after annexing parts of Punjab, the British recognized Maharaja Gulab Singh, the strongest ruler north of the Sutlej River, as the ruler of Jammu and Kashmir. But for this, he had to pay the sum of Rs 75 lakhs in cash; this payment was legal as the maharaja was a former vassal of the Sikh Empire and was partly responsible for its treaty obligations. Maharaja Gulab Singh was thus credited as the founder of Jammu and Kashmir. After his descendant Maharaja Hari Singh, the last ruler of Jammu and Kashmir, signed the Instrument of Accession in 1947, Jammu became a part of the Union of India.

Jammu enjoys the status of an administrative division within the state of Jammu and Kashmir. Jammu City, the largest city in the region, is the winter capital of Jammu and Kashmir. The majority of Jammu's 2.7 million population practises Hinduism, while Islam and Sikhism enjoy a strong cultural heritage in the region. Due to relatively better infrastructure, Jammu has emerged as the main economic centre of the state.

1.5 CULTURE

Jammu region is home to several ethnic communities which follow traditional lifestyles but with distinctive cultures of their own. Among these communities, the Dogras constitute the dominant group. They are mainly concentrated on the outer hills and outer plain zones covering Kathua, Udhampur, and Jammu districts and the lower parts of Rajouri District. A martial community by tradition, their folklore centres on eulogies for war heroes, both legendary and historical. Even the region's architectural heritage, comprising elaborate castles and hilltop fortifications that are visible everywhere, bespeak the community's long-drawn preoccupation with battles and ruling of distant lands. Yet the region's history is not completely bereft of traditions of art and culture. Thus, while the troops fought battles in distant areas, the royalty and the nobility nurtured art and culture. The Pahari miniature paintings, which have justly become famous throughout India, are the finest examples of their artistic achievements.

The second largest ethnic group of the region is formed by the Gujjars, a semi-nomadic people living along the hill slopes of Doda and Rajouri districts; in Poonch, they also dominate the main valleys. Some of them have settled down to agriculture, but the majority are primarily herdsmen. They cultivate maize along the slopes of the mountains but only as a secondary occupation.

1.6 ARRANGEMENT TO CROSS CHENAB

It was a real challenge for the people of Chamb, Jouri, and Akhnoor to cross the river Chenab and contact the people of Jammu as well as the rest of India since civilization. In 1922 a historical event was celebrated on the bank of this river in the form of Raj Tilak for the first king of the Dogra dynasty, Maharaja Gulab Singh.

Analyzing the difficulties of the people, Maharaja Gulab Singh decided to construct a steel bridge and to enable motorists to cross the river Chenab as well as to open for trading business among the people. Thereafter, a three-span (1 × 138 m + 2 × 35 m) steel-girder through-type bridge with a total length of 208 m at 25.25 km to cross the river Chenab was constructed on 1932–1934. The steel bridge was constructed by M/S Jessop & Co., and it was named as Yuvraj Karan Singh Bridge. Hence, the Akhnoor, Jouria, Palanwalla, Chamba, etc. areas became accessible to Jammu by surface communication for restricted one-way class of vehicles.

The passage of items and the increasing population and the day-by-day requirements for shifting heavier loads required higher load classifications of vehicles/equipment/plants by the army authority to IB/LoC after the birth of Pakistan; thereafter, the existing loading capacity of the steel bridge was upgraded to Class 24.

The government of India desired to improve the road and bridge up to the IB/LOC, considering its strategic importance after the occurrence of the famous battle of Chamba–Jouria in 1965 and the Indo-Pakistan War of 1971. Therefore, the state government of J & K decided to hand over this road to BRO in 1971, along with the Akhnoor steel bridge.

2.1 BRIEF ON STEEL BRIDGE

The three-span through-type bridge at 25.25 km on Jammu Akhnoor road has been constructed with steel girder over brick piers and abutment between 1932–1934. The bridge was designed by the Indian Railways for a single lane 4.5 m in width, and it was constructed by M/S Jessop and Co. under the guidance and supervision of railway engineers. The single largest span of this bridge is 138 m; whereas, the other two equally span 35 m each. The total length of this bridge measured after construction is 208 m. The span arrangement of this bridge is shown below:

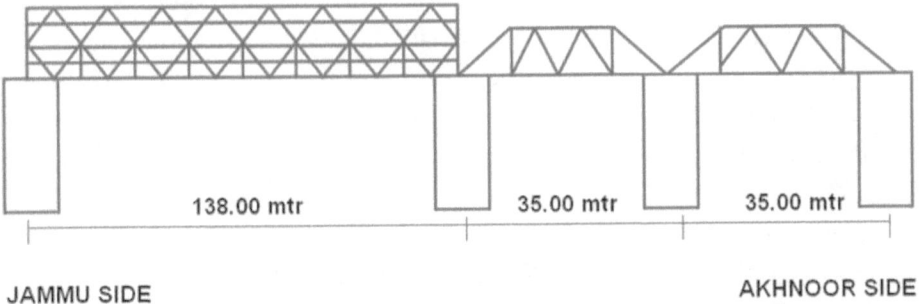

JAMMU SIDE   AKHNOOR SIDE

Fig. 2: Span arrangement of steel. Bridge

The bridge was initially constructed with timber decking for light traffic; later, the timber decking was changed, and the 24R loading class was upgraded. The demerit of this bridge was observed with the passage of time and the increasing population and the requirement of shifting heavier loads, which necessitated higher load classifications for vehicles/equipment/ plants by the army to IB/LoC after the birth of Pakistan. This required the construction of the permanent bridge over the river Chenab near the existing steel bridge. These strategic needs necessitated the construction of a new double-lane bridge with Class 70R loading as per the current loading standards of the IRC.

The government of India desired to improve the road and bridge to improve accessibility up to IB/LoC, considering its strategic importance after the famous battle of Chamb–Jouria in 1965 and the Indo-Pakistani War in 1971. Therefore, the state government of J & K decided to hand over this road to BRO in 1971, along with the Akhnoor steel bridge.

In September 1971, the road, including the existing Akhnoor bridge, was taken over by the BRO, the premier road construction wing of the government of India. Immediately thereafter, a detailed ground survey for the improvement of the road to a double-lane road and the construction of a permanent bridge over Chenab was carried out.

2.2 SERIOUS PROBLEMS IN THE STEEL BRIDGE

On 10 September 1992, the superstructure, 138 m span of the bridge, was washed away in a flash flood, and the entire area up to the border became cut off. There were a huge cry from the civilian and army populations. There was huge pressure from every part of the country for the restoration of this bridge and for the immediate construction of a permanent concrete bridge.

Reconstruction of the bridge was started by Northern railway on December 1992 after a detailed investigation over a raised abutment and pier by 1.2 m, and it was completed within 16 months (i.e. by 13 April 1994) with a total cost of Rs 5.78 crores for Class 24R loading and was opened for traffic.

The undermentioned parameters have been adopted for the reconstruction of this bridge:

- ➢ It is a steel-girder through-type superstructure.
- ➢ The pier and abutments were raised 1.2 m in height.
- ➢ Span arrangement is 1 × 138 m + 2 × 35 m.
- ➢ Callender–Hamilton components were used.
- ➢ Joints are codigon joints in vertical and bracing members.
- ➢ Carriage way width was kept at 4.4 m.
- ➢ Decking was steel plate covered with mastic asphalt.

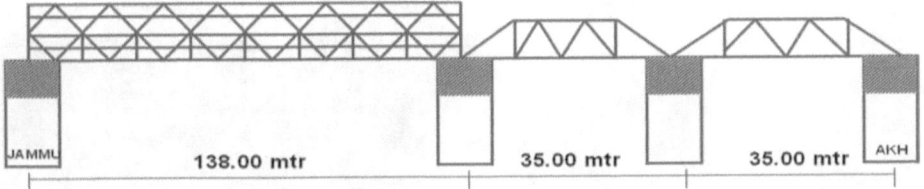

Fig. 3: View of the steel bridge on 1.2 m raised abutment and piers.

➢ Four pot bearings were provided for smooth movement.
➢ Structure was designed for 270 mm hogging.

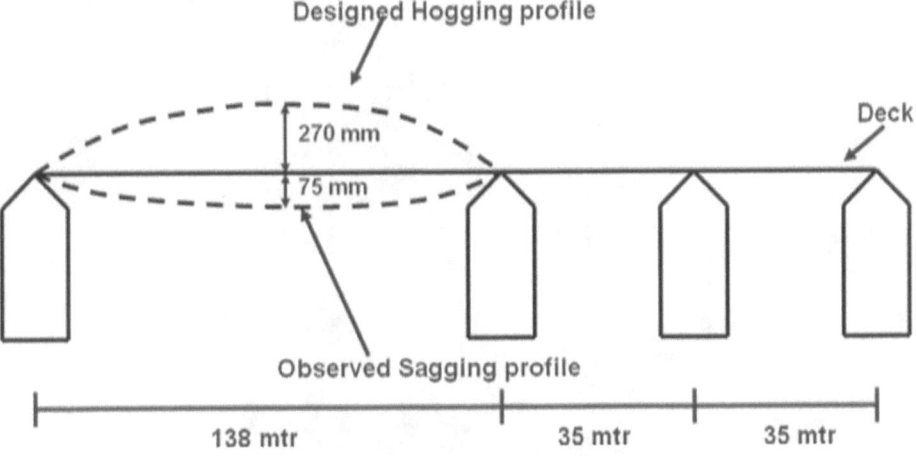

Fig. 4: Sagging profile of the steel bridge after launching.

➢ On completion of restoration, the largest span had observed 75 mm sagging in place of hogging at the time of launching had been kept with the structure. Hence, it is a clear-cut design failure, but the same could not be pointed out by the BRO. Therefore, 32 prestressed cables, 16 on each side, were provided with the structure and were stressed up to the designed load.

Fig. 5: Repair of the bridge after damage of prestressed cable.

The bridge was kept under close observation and under guard by the army authority. Only the vehicles strictly up to the 24R loading class were allowed to cross by the army sentry.

On March 1999, some deficiencies like sagging, some missing bolt/nuts, distortion of bearings of bearing bolts and plates, and damage of surface were observed, and a higher formation was reported.

M/S Consulting Engineering Services (CES) Pvt Ltd, India, was requested to inspect and suggest remedial measures. In response, M/S CES Pvt Ltd suggested a condition survey of the bridge's safety, serviceability, and durability without describing the technical specifications of the proposed rehabilitation work, execution methodology of proposed works, etc.

Based on the preliminary report and actual requirement on the ground, the prestressed cable was replaced in 2001, and load classification was restricted to Class 18R from Class 24R on June 2002. Further, no deformation was observed or progressed.

The maintenance team was deployed round the clock to watch the behaviour and tighten the nuts and bolts once every 24 hrs to set it right or back to the original position.

2.3 EARTHQUAKE EFFECT ON BRIDGE

There was a high-intensity earthquake observed on 8 October 2005 in the surrounding area. Due to the earthquake and the wear and tear effect, the three prestressed cables were found damaged on 9 October 2005 early morning. On the same day, a BOO was detailed to inspect the bridge properly. The BOO recommended the replacement of the 29 balance cables and the damaged prestressed cables up to the designed stress immediately. The BOO further advised to reduce the load classification to Class 9 till the replacement of the 32 prestressed cables. All the 32 cables were replaced and stressed up to the designed strain by the M/S D2S Infrastructures at the cost of Rs 9.99 lakes in record time, and the bridge was opened for vehicles with Class 18R loading on 31 January 2006.

2.4 SPECIAL ATTENTION

The 24 hrs electricity supply was restored with traffic-controlled device from the maintenance team. The case for a separate maintenance grant was taken up by the competent authority. The maintenance grant was finally approved by the competent authority after proper analysis and actual requirements on the ground were ascertained.

3.1 THE CASE FOR PERMANENT BRIDGE SCHEME

It is a fact that scarcity is the mother of invention. The day-to-day problems of the steel bridge, the restriction on loads, the single-lane capacity, and the very high maintenance cost compounded the immediate necessity for the construction of a permanent bridge.

The famous battle of Chamb–Jouria in 1965, the Indo-Pakistani War in 1971, and Kargil War in 1999 have shown the importance of the construction of a permanent bridge over Chenab near the existing site of the steel bridge. The importance of the construction of a concrete bridge with a higher load classification pushed the government of India for immediate construction of a permanent bridge.

In September 1971, the road, including the steel bridge, was taken over by the BRO (the premier road construction wing of the government of India). Immediately thereafter, a detailed ground survey for the construction of a permanent bridge over Chenab was carried out. The most suitable site for the construction of a permanent RCC bridge was finally selected, and subsoil investigation on the ground was carried out by the Geological Survey of India and was completed in 1975. After the geological investigation and classification of the ground strata, a scheme for a bridge 231 m in length and with five spans ($3 \times 46 + 2 \times 46.5$) at 800 upstream of the existing steel bridge was proposed and duly considered the site nearer to IB/LoC and made an alternative superstructure by launching a steel bridge in between the pier or abutment to the pier if the superstructure is damaged in war without any time lack. The scheme had been finally approved by the competent authority of the government of India.

After completing the formalities, the contract for the construction of the 231 m long five-span RCC bridge was awarded by the BRO at a cost of Rs 59 lakhs to M/s National Building Construction Corporation (NBCC) in 1978 for completion in three years—i.e. up to 26 December 1981.

The construction of the bridge was commenced on the ground in 1978 by the NBCC but could not be completed even after the extended PDC up to 28 Apr 1984. Only 25.02% of the work was completed by NBCC, but they could not carry out well-sinking in the bouldery strata. They ultimately abandoned the work and absconded from the site, abandoning the balance (74.08%) of the awarded work. The sum of Rs. 24.28 lakhs had been paid to NBCC, and a joint inventory for supply/machinery was prepared by the board of officers. The cost for the stores left by NBCC at the site was Rs. 5,17,761 supply. The approved scheme and the work executed by NBCC is depicted below:

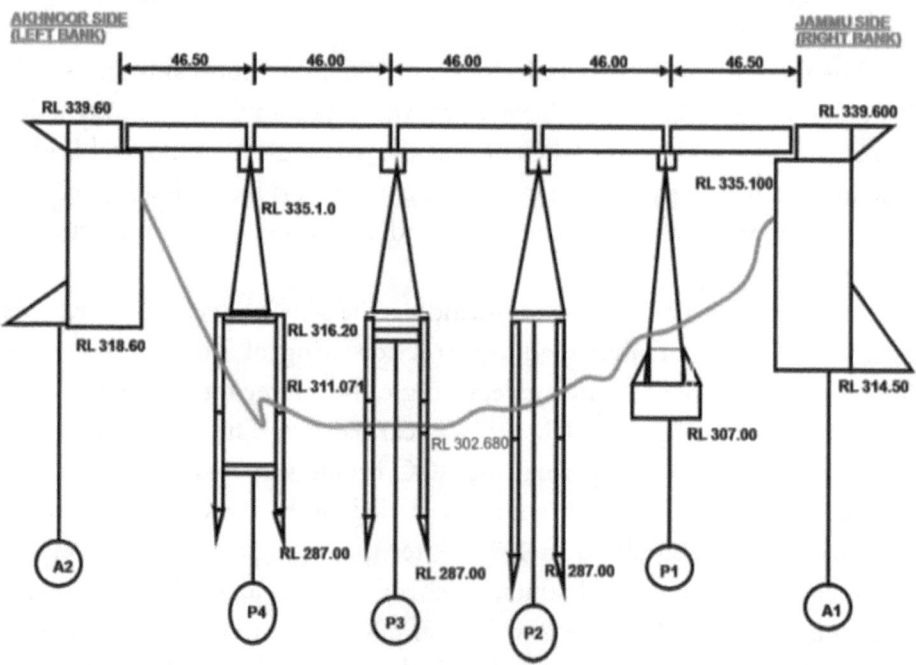

Fig. 6: Approved bridge scheme.

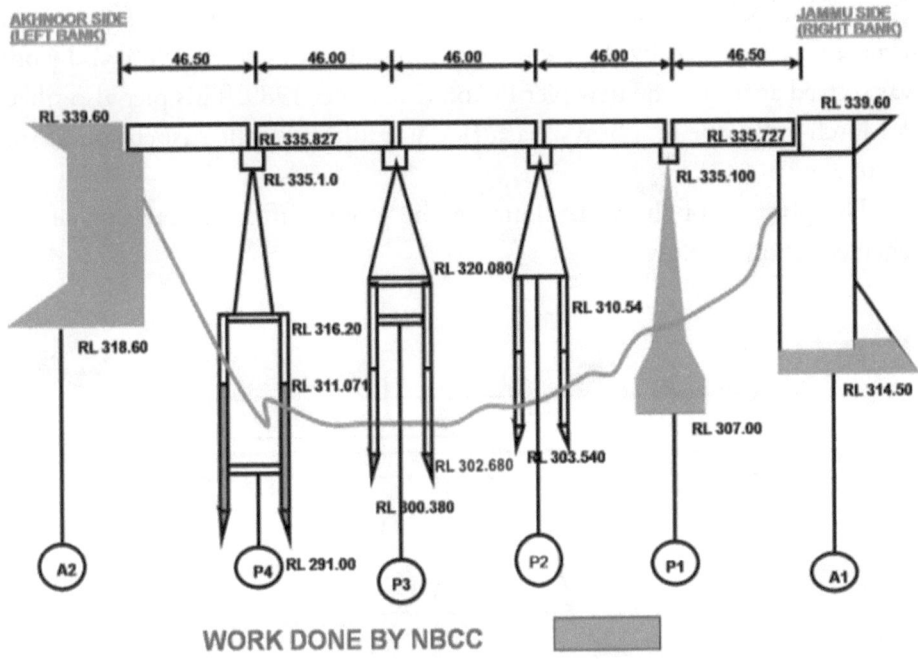

Fig. 7: Work executed by NBCC.

The contract of NBCC was finally cancelled as per CA provision, and BRO decided to carry out the balance work of 74.98% through other contract agencies. All the formalities were carried out, and the work was awarded to M/s Banka India Ltd on 23 December 1986 for a sum of Rs 212 lakhs. The cost of the stores received from NBCC, which was Rs 5,17,761, was further issued to M/s Banka India Ltd under Schedule B.

The work was resumed at the site by M/s Banka India Ltd on Jan 1987. The work on the abutment on both sides and pier P2 and casting of girders had been initially taken up. The work suffered a setback when pier P2 tilted in Mar 1988 by huge floods. However, the works on other wells and abutments were continued, but in the highest flood in July 1989, pier P3 also got tilted beyond recoverability. Banka India Ltd could not carry out well-sinking in the bouldery strata and tilted piers could not be corrected.

M/s Banka India Ltd projected a case for the revision of the original scheme due to difficulties faced on the ground. BRO analyzed and decided to revise the scheme. M/s Banka India Ltd was directed to submit the proposal.

The original scheme was revised by introducing a new pier PC. The revised scheme was approved by the competent authority in October 1989, and work was started again on the new pier PC on December 1989. This pier also tilted in March 1990 due to a heavy flash flood, resulting in the suspension of the entire work.

The position of the work done on the ground after the revision of the scheme is shown below:

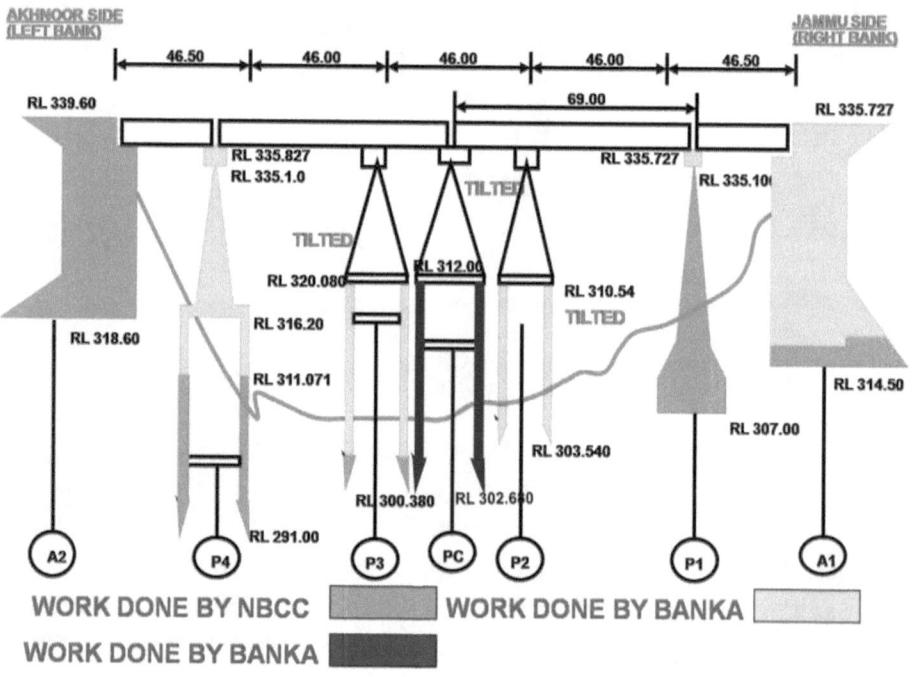

Fig. 8: Revised original scheme; work executed by Banka and damaged.

M/s Banka India Ltd revised the scheme and submitted it in March 1990, duly adding two new piers, P11 and P44, by abandoning PC pier. This proposal was also approved in October 1993 without properly addressing the cause of repeated tilting of the well in the river. The new scheme was finally approved.

The work was commenced on the new pier P11 on December 1993. This pier also got tilted on June 1994, and the contractor, M/s Banka India

Ltd, stopped all work at the site. M/s Banka India Ltd also failed in their attempt at well-sinking, and the contractor ran away and abandoned the site. A sum of Rs 128.58 lakh in payment against RAR so far had been made to M/s Banka India Ltd. A dispute was raised on the due payments between the contractor and the department. The contractor went to arbitration. The arbitrator awarded an amount of Rs 24.55 lakhs in favour of the contractor and Rs. Nil to the department on 9 May 1998. On scrutinizing the award, the department did not accept the judgment and appealed to the J & K high court. Finally, the contract was cancelled by the accepting officer on 1 January 2003. The proper inventory of stores/vehicles/plant equipment left at the site was accounted for by BOO. The depreciated cost of all items was calculated by the BOO to be Rs 3,57,800.

The final status of the work executed on the ground by both contractors is shown below with a view of the site:

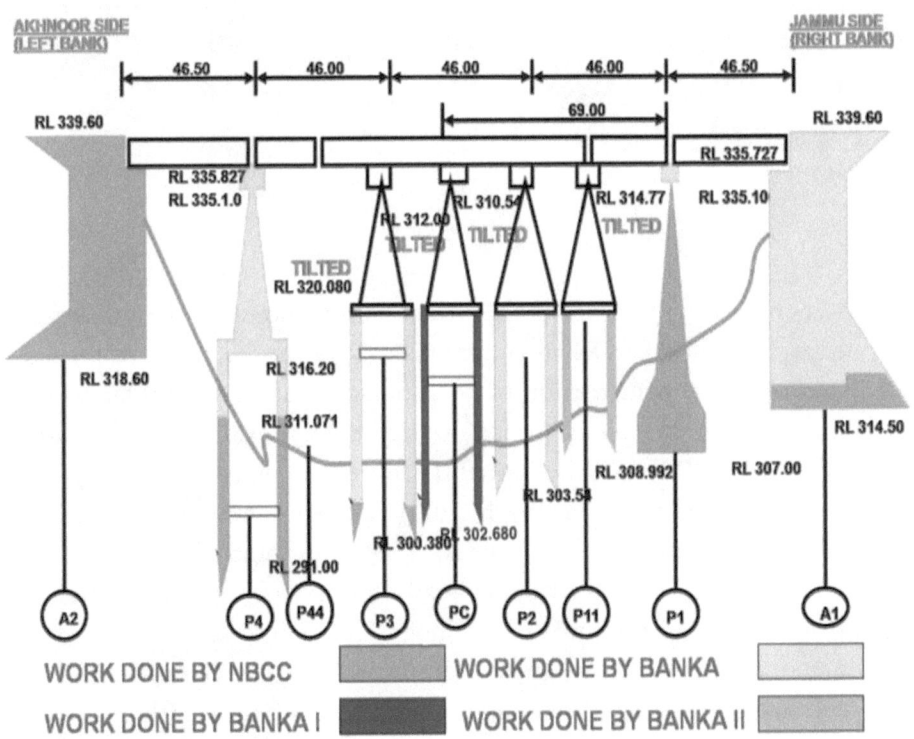

Fig. 9: Position of work on the ground.

Fig. 10: View of abandoned site.

As soon as construction of the bridge commenced on the ground on December 1978, the BRO had simultaneously processed the case for land acquisition and the construction of an approach road to the bridge so that bridge, along with the approach road, would be made available to the nation by 26 December 1981. All the formalities were completed for the approach road, which was 7.518 km in length, and work was commenced from both ends—i.e. Akhnoor and Jammu sides. The take point of approach road is 23.3 km on the Jammu–Akhnoor road and meeting at 29.2 km on the same road, bypassing the town of Akhnoor. The length of 2.91 km on the Akhnoor side and 1.073 km on the Jammu side (DL specification) of the approach road had already been completed. The 2.99 km length on Jammu side, formation cutting/filling, and GSBC partly had also been completed by 1992. Therefore, only a balance of 0.545 km in length remains for the formation and surfacing work.

The position of the approach road is depicted below:

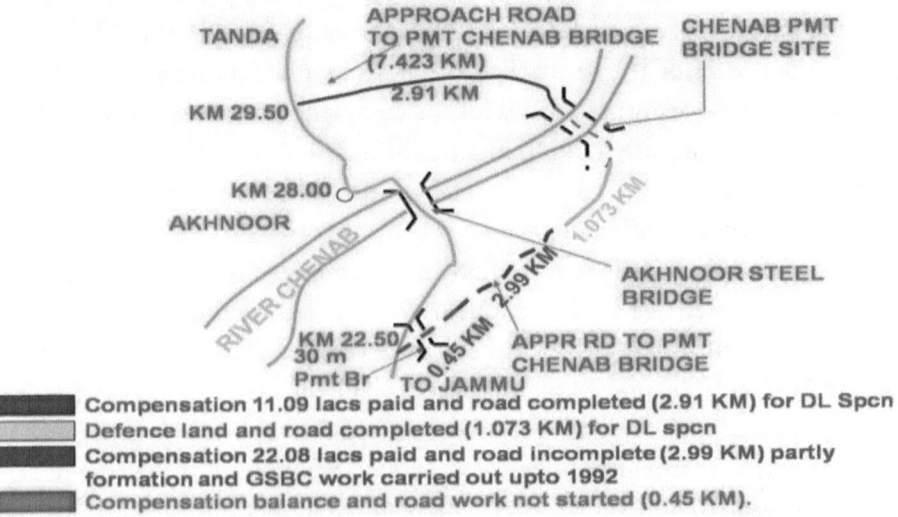

Fig. 11: Complete view of the approach road.

3.2 NEW EFFORTS TO RESTART THE BRIDGE WORK

On cancellation of the contract, a new attempt was made for the construction of the bridge at the selected site, and tenders were called for design and construction of a permanent bridge at the existing abandoned site on September 2003. The tender was opened on 12 December 2003, and it was found that M/s S. P. Singla Constructions Pvt Ltd had quoted the lowest rate of Rs 16.39 crores. The final proposal and design was asked from M/s S. P. Singla Constructions Pvt Ltd by BRO to ascertain the viability of the scheme, but they had failed to submit the details in time. Therefore, the tender action had been cancelled, and hope for construction of the bridge petered out.

3.3 ATTEMPT WITH A NEW CONCEPT

The general public (J & K) and the Indian Army demanded for the immediate completion of the concrete bridge over river Chenab near

Akhnoor on Jammu. The Akhnoor–Poonch Road in the state of J & K lying incomplete for more than 24 years prompted all concerned in the decision-making process to find ways and means to complete the permanent bridge to meet the strategic needs of the nation and to overcome the traffic bottlenecks faced by the users of this road. The tremendous pressure from the political leadership and the army authorities for the completion of the bridge forced the BRO to decide on many bold, dynamic, brilliant, and viable administrative and technical decisions. The situation demanded some viable solution, so they had to think unconventionally.

The author was assigned the job of commander of the task force 13 BRTF and reported in May 2005. On arrival and during the course of taking charge over of the task force, it has been observed that the construction of the Chenab Bridge is an unsolvable problem for all the civil engineers of BRO. Therefore, I had taken a challenge, and I started hunting for the entire record of the problems that had occurred in the past. A number of times the bridge site was inspected, and the ground data were analyzed as well as facts from the past from historical records of J & K. While reading a historical book on J & K, it was noticed that writer had already forecasted in the fifteen century that no permanent structure is advisable for construction in the river Chenab. It has also been observed that the riverbed strata are very hard conglomerate soil, which are unsuitable for well-sinking.

The director general of Border Roads (DGBR), Lt. Gen. K. S. Rao, SM, had made a programme to visit the 13 BRTF sector in August–September 2005, where the author had been posted. Therefore, the author decided to propose a new scheme for Chenab Bridge, duly addressing the reason for failure and providing a solution to permanently resolve the issue. The DGBR finally visited on 31 August 2005; four types of various proposals were discussed during the briefing, and finally, one proposal was recommended based on the merit best suited for site. These proposals are being elaborated in succeeding paragraphs.

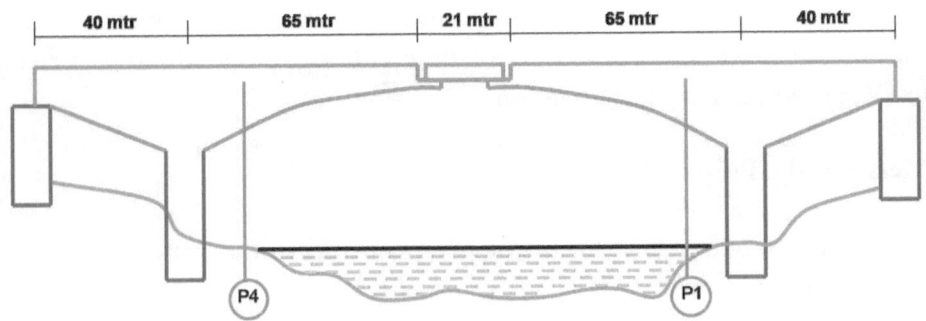

Fig. 12: Brief of proposals.

Fig. 13: Bridge scheme of 231 m.

Proposal 1: Prestressed, Balanced Concrete Cantilever Bridge

Keeping in view the site's condition, 231 m length of approved bridge scheme, high velocity of the water current, conglomerate soil strata and the difficulty in sinking the well on the ground without pneumatic method, and the hostile behaviour of the river, the cantilever bridge proposal was projected to DGBR without any apprehension.

Merits of the Proposal

- ✓ The acquired land for use of 7.518 km road will be fully utilized for construction.
- ✓ The cost of the construction and maintenance of the cantilever bridge will be cheaper than the construction of a new larger-span bridge, cable-stayed bridge, etc.
- ✓ The recurring cost of maintenance for the cantilever bridge is less than that of any other bridge.
- ✓ There are no chances for flood damages to the foundation and substructure during construction as both piers are posed to be constructed on dry and firm conglomerate soil strata.
- ✓ Two piers already constructed will not be utilized but will act as a protection of other piers from erosion or flood effect.
- ✓ Excavation will be easy as foundation is open.
- ✓ Construction will be speedy and without any difficulties for both sides of the bridge.

Demerits of the Proposal

- ✗ Two abutments and nine concrete girders casted and lying on the ground will not be utilized.

Proposal 2 and 3

The construction of a 231 m length cable-stayed bridge (proposal 2) and the construction of a 231 m length steel girder bridge (proposal 3) at the same site have been discussed, but both proposals were found to have a number of disadvantages.

Proposal 4

The construction of 400 m length RCC Pmt bridge at a new site was also discussed, and it was observed that this new proposal might take much time as it involved the process of land acquisition and other formalities—bridge design and preparation of a new site; therefore, the proposal did not appear practicable, feasible, or economical.

3.4 RECOMMENDATION

Considering the merits and demerits of each proposal, economical feasibility, availability of technology, construction that is speedy and free from obstruction, full utilization of the approach road's length, as well as minimum wastage of expenditure added to what was already incurred in this scheme, the first proposal was recommended by the author.

The complete scheme of the first proposal was again discussed at the site on 1 September 2005 by the Lt. Gen. K. S. Rao, SM, while he visited the site personally to ascertain and verify all the facts. Ultimately, he accepted this scheme in principle, with some modification to the length/span.

Fig. 14: Visit to the bridge site.

Fig. 15: Discussion of the proposed scheme at the site.

3.5 FINALIZED PROPOSAL

The proposal was examined in detail by the competent authorities, and finally, a 280 m long continuous cantilever bridge having a single main span of 160 m that necessitates keeping the pier wells inside the bank and any from the water line was approved. All old structures constructed on the site, whether complete or incomplete, were abandoned. This kept the foundations away from the water's edge, and overlapping with the foundation of the old existing piers, P1 and P4, was also avoided. This scheme also facilitated the construction of the foundations on dry land, free from any obstruction. The finalized bridge scheme is shown below:

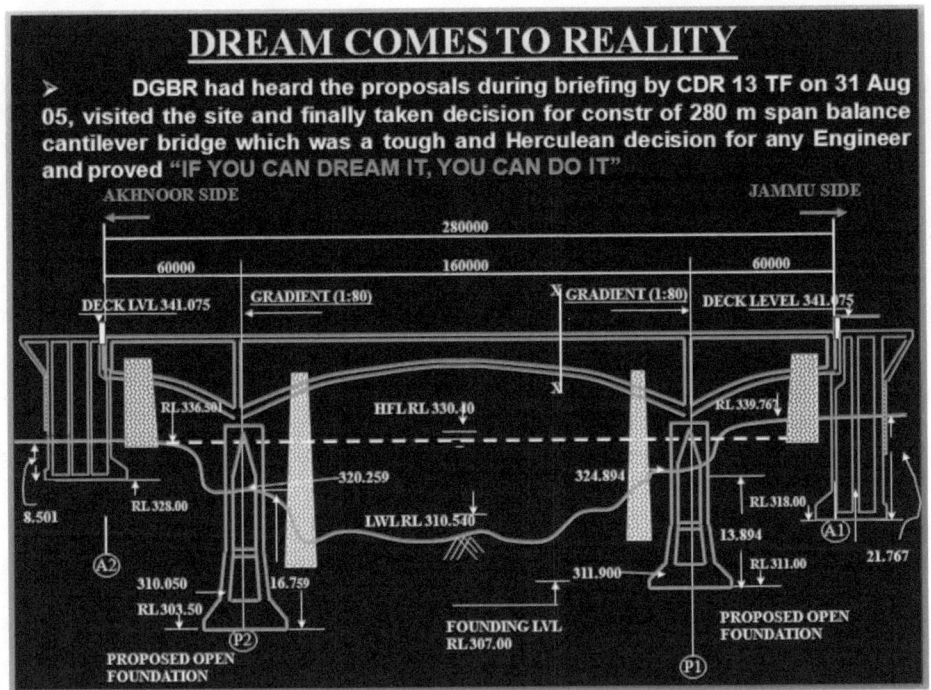

Fig. 16: View of the newly approved bridge scheme.

This culminated in a bold, dynamic, and brilliant administrative and technical decision, giving fillip to the hope for the construction of a permanent bridge over the river Chenab at Akhnoor. The boldness of the administrative and technical decisions are described as follows:

Technical Decision

➤ The new proposal eliminated all foundations in the water.
➤ All the earlier contractors had failed because of problems encountered in the construction of the foundations, with the hard conglomerate soil strata causing difficulties in well-sinking, and the turbulent water current resulted in the tilting of the wells. Hence, the proposal was to go for open foundation for the bridge and to bridge a river gap of about 130 m without any support. Thus, a main single span of 160 m, with two ends spanning 60 m each, was selected to keep the foundation edge well away from the water line.

> As a collar to the above decision, all the foundations remaining on the ground and in the river were abandoned. The PSC girders lying on the ground towards the Jammu side embankment were also abandoned.

Administrative Decision

> The technical recommendations by HQ DGBR (stated above) were accepted by the government BRDB secretary, resulting in the writing off all the past expenditures made in the construction of the bridge.
> It was also decided by the government to call for tenders by invitation only, and accordingly, 15 preselected firms approved by the government were invited to bid for the work.
> Very stiff conditions of payment for executing the works were imposed on the NIT, which were not negotiable. They are reproduced below:

- No running or advance payment for materials or for machinery were allowed till the levels of the piers crossed the designed HFL.
- No running payments were admissible till the piers P1 and P2 crossed the HFL. No payments for abutments were to be made until both the piers crossed the HFL RL of 330.4 m.

The time of completion, 20 months, was kept sacrosanct, and liquated damages, 10% of the CA amount, were liable for any delay in the period of completion.

The interim payment schedule was fixed, where 8% (i.e. more than Rs 160 lakhs of the CA amount) was kept for load-testing among others. The exact conditions of the NIT/tender are reproduced below:

Billing Schedule for Advance on Account Payments (S1 page 61 and 62 of Tender)

Tender Clause 25.1 (a) The pricing for payments of advances on account of work done and of materials delivered will be as given hereunder:-

S/No	Item of work	Expected completion period/date from the date of commencement	Percentage payable
((i)	On submission and approval of detailed design and drawing	06th Month	2%
(ii)	On completion of Pier marked P1	12th Month	8%
(iii)	On completion of pier marked P2	12th Month	8%
(iv)	On completion of abutment marked A1	10th month	4%
(v)	On completion of abutment marked A2	10th Month	4%
(vi)	On installation of bearing	12th month	1%
(vii)	On completion of super structure	14th month	
	Home bank span of 60 M		11%
	Far bank spans of 60 M		11%
(viii)	The centre span	16th month	25%
(ix)	Casting of deck slab	17th month	14%
(x)	Hand rails	18th month	2%

S/No	Item of work	Expected completion period/date from the date of commencement	Percentage payable
(xi)	Completion of load testing	19th month	8%
(xii)	Site clearance	During 20th month	2%
	Total	20th month	100%

Note:- The contractor shall note that the compensation for delay, if any, in completion of work shall be levied at the rate of 1% per week of the amount of percentage indicated above for the delayed items as a whole (including the completed portion also) of works against which separate completion period has been given. The maximum amount of compensation (LD) shall be limited to 10% of the amount of percentage payable indicated above. However, the total amount of compensation shall be limited to 10% of the contract amount.

(2.5.1(b)) In case the tendered is submitting his own new proposal and the scheme other than the departmental proposal he shall submit the schedule for running payment also for approval of the Accepting Officer along with his tender.

(25.1.1.) Advance on payment schedule as given above will be governed by the order of priority of construction as indicated above i.e. contractors will complete both the piers up to HFL level before undertaking construction of abutments. In case, however, the contractor desires to undertake the work of abutments prior to that of piers or simultaneously he shall be at liberty to do so with expressed condition that any advance payment on account of any of the abutment works will be released as per above percentage only, when progress of works on both the piers has reached at least up to HFL level. No advance on payment for material at site for superstructure work will be made till the piers are completed up to at least HFL level.

(25.1.2) Once the construction of piers (abutments in case of single span cantilever construction) reaches HFL level, advance on account payment as clause 24(a) above may be claimed by the contractor as per work done and material brought at site at Pro-rata to the percentage indicated against each individual items(s) of works.

(25.1.4) In case of single span cantilever construction from both ends, no advance on account payment, either for abutments or for materials brought at site, will be made till construction of both the abutments reach up to HFL, Thereafter, the advance on account payments shall be governed by cause 25.1.2.above.

(25.1.5) No payments shall be made against the perishable material. The decision of Engineer in charge, as to the perish ability of any item shall be final and binding.

(25.1.6) the advance payments against the material shall be adjusted fully from the next payment of advance against the works executed. If, however, the entire advance is not adjusted in the first RAR, it will be adjusted in the next RAR.

The insurance of the construction covering the entire period of the work was borne by the contractor.

Based on the key technical and administrative decisions above, the scheme was finally approved, and the technical price bids were called for the prestressed continuous cantilever bridge with a total length of 280 m (60 m + 160 m + 60 m) as shown at Fig. 16.

Accordingly, bids were received and analyzed. On analysis, it was observed that D2S Infrastructures Pvt Ltd had quoted the lowest and workable rate. Therefore, the work was awarded to M/s D2S Infrastructures Pvt Ltd at their quoted rates of Rs 21.06 crores.

CHAPTER

4

4.1 DESIGN PHILOSOPHY OF THE BRIDGE

The tender was based on the successful contractor's own design. The successful contractor utilized the service of the design consultant Shri R. B. Singh.

4.2 DESIGN CONCEPT

As per the contract agreement, the design of the concrete bridge was based on IRC standards applicable and was revised from time to time as conceptualized by Shri R. B. Singh of Distinct Planning and Design Consultant Pvt Ltd, a brilliant and dedicated young engineer with a great sense of aesthetics.

The existing structure and landscape of the site were fully taken care of so that obstruction due to the piers would be negligible. The fixing of the foundation size had also been given due weight age so that the edge on the water site was maintained. All the components of the bridge were designed to work together and complement each other aesthetically. It was further ensured that the substructure and the superstructure box had the same width (i.e. 500 m) in cross section so that the slenderness and sleekness of the structure is maintained. This was done to enhance its aesthetics, which in fact is a visual delight when actually seen.

4.3 LOADS

The characteristic values of Pm (x, t) have been used for prestressing. For the erection stage, relevant values of t have been used. For the finished structure, are used values of t just after completion and after 70 years? The variable loads considered are:

(a) **Traffic Load:** One lane of Class 70R or two lanes of Class A, according to IRC: 6-2000, clause 207.1.

(b) **Footway Loading:** According to IRC: 6-2000, clause 209.4:　P $= (P' - 260 + 480 / L (16.5 - W / 15,$ P' $= 500$ kg/m², where $L =$ length in metres of relevant part(s) of influence lines, $W =$ width of the footway.

(c) **Braking Force:** 20% of the load of the first train and 10% of the load of succeeding trains; train loads in one lane only being considered.

(d) **Water Current:** Horizontal forces from the water current are based on a mean velocity of 3.5 m/s.

(e) **Temperature:** For concrete structures, uniform temperature change in the structure $\pm$ 25 °C; non-uniform temperature distribution in as per positive and reverse temperature differences as per clause 218.3 of IRC: 6-2000.

(f) **Erection Loads:** Traveller-type top truss, weight age 650 kg, point of application from support tip of cantilever, imbalance due to own weight (dimensional tolerance) 5%.

(g) **Random Loading:** (equipment and personnel) 0.5 kN/m².

(h) **Vertical Wind Pressure:** 0.55 kN/m².

(i) **Concentrated Load at Tip of Cantilever:** 50 kN.

(j) **Bearing Friction:** Sliding bearing (Teflon on stainless steel 5% of the actual characteristic bearing load).

(k) **Accidental Loads from Earthquake:** The site is in zone V and zone factor is 0.36.

4.4 DESIGNS OF VARIOUS COMPONENTS

Superstructure

The superstructure was constructed progressively from piers. The PSC box depth varied from 9.75 m at the root to 3.4 m at the centre of the span 80 m away from the piers. Span depth ratio of mid span at the root was 16.4 m, and at mid span, it was 47 m. The width of the box at the deck level was 12 m and comprised of a carriageway width catering to two-lane traffic with 1.5 m footpath on either side separated by steel crash barriers. The soffit profile provided was parabolic.

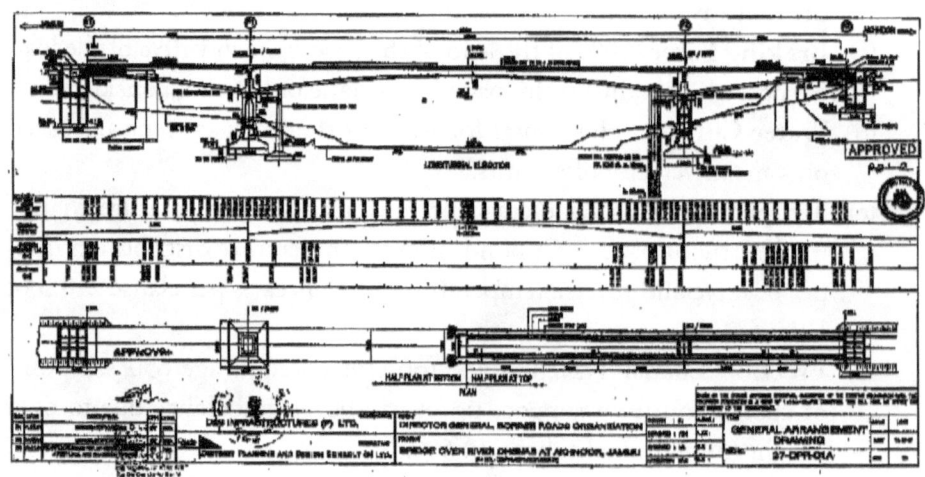

Fig. 17: View of soffit profile.

Prestressed Cables

For prestressing, 62 × 19 T 15 cables at the junction of the web and deck slab with suitable reinforced blisters were provided. Thirty-one of prestressing cables 19 T 15 on either side of the box were anchored at the junction of the web and deck slab.

Provision for future prestressing were kept by providing five 110 diameter holes in the diaphragm at the pier head.

Blisters

Blisters have been provided either in the web or essentially at the junction of the web with the deck or the soffit slab.

At the pier locations, a diaphragm wall was provided, which was 1,200 mm, with the bottom part above the soffit as a solid block (size 4 m × 4 m) suitably reinforced as per the design.

Substructure and Foundation

The abutments consisted of a hollow box, and piers had a hollow rectangular section on raft foundations. The pier section is tapered near the foundation to a height of 4.5 m for effective dispersion of the huge load on the foundation area.

Abutments

The box size of A1 was 12 m × 10.5 m and suitably divided into nine cells. The box size of A2 was 12 m × 7.6 and divided into six cells. The wall thickness was 750 mm in the front; the back walls were 500 mm, and the side walls were 450 mm.

Piers P1 and P2

The hollow piers' section size was 6.6 m × 5 m outer dimension. The raft foundation size for piers was 14.25 m × 17 m.

Stability during Construction

During construction, stability cables at the abutments were provided to prevent uplift at abutment along with bents near the piers (located at 12–14 m from the piers towards the earth's surface).

5.1 EXECUTION METHODOLOGY

Before starting the execution, the construction methodology was finalized and divided into various stages, which are depicted below:

Stage 0	Dismantling of the top portion of the existing abutment and existing pier (EP1 and EP2).
Stage 1	Ensure timely completion of the project; two independent sets of concrete production, transport, and placing equipment as well as cantilever construction gantries were provide (one for each bank).
Stage 2	Cast pier head segment 1 on bearings and temporary blocks for stability (supported on pier cap).
Stage 2a	Erect staging on either side.
Stage 3–5	Cast segments 2 to 5 simultaneously on either side of the pier head, accounting for camber and prestress progressively.
Stage 5a	Provide support on elastomeric bearing on temporary bent (bent 1/bent 2) at 11 m from the pier centre in span 1/span 3 and on existing pier (EP1/EP2) 11 m from pier centre in span 2. Contact of super structure with bearing on temporary support (bent/existing pier) shall be ensured after prestressing through jacking or otherwise.
Stage 5b	Erect traveller on river side and equivalent weight (by trolley on rail or on land side).

Stage 6–18	Cast segment 6 to 18 simultaneously on either side of pier head, accounting for camber and pre stress progressively (segments were to be cast on staging in span 1 and 3 by shifting staging progressively and by traveller in span 2).
Stage 18a	Cast abutment diaphragm and box projections.
Stage 18b	Place temporary supports on abutment and prestress stability cables for holding down. (Provisions to be kept in abutment wall. Rotation and movement at abutment on long, long axis not to be restrained by providing sleeve around the cable to allow flexibility). Remove trolley on land side.
Stage 18c	Remove support on existing piers, EP1 and EP2.
Stage 18d	Remove temporary bent—bent 1 and bent 2.
Stage 18e	Prestressed continuity cable numbers 201–203, 205–207 in span 1 and span 3. Fill with PCC in segments near abutment, stress cable no. 204; cast U box at diaphragm and fill with PCC.
Stage 19–22	Casting and prestressing of segments 19 to 22 in span 2 will progress with camber correction.
Stage 22a	Move traveller systematically close to the centre line of the pier, and dismantle.
Stage 22b	Suspend form for closure segment for mid span (central span), and erect form work for closure span.
Stage 23	Cast closure segment and stress cable numbers 101, 103.
Stage 23a	De-stress stability cables, and stress the remaining bottom continuity cables in span 2. Cast balance abutment part, deck slab, and side walls, and install top and bottom bearings of abutment.
Stage 23b	Complete construction—crash barrier, railing, expansion joint footway, and wearing coat.

5.2 LAYING THE FOUNDATION STONE

Fig. 18: Laying the foundation stone.

Fig. 19: Chief minister discussing the scheme.

The contract for the design and execution was accepted after the completion of all formalities with M/s D2S Infrastructures Pvt Ltd on 27 April 2006. The foundation stone was laid by Hon'ble Chief Minister Janab Ghulam Nabi Azad on 28 April 2006.

5.3 EXECUTION PROGRAMME

To ensure timely completion of the project, a detailed programme had been prepared by the author as commander of task force, along with the contractor, project manager, and designer on 3 May 2006. Important decisions as mentioned below were also taken so that the work would be completed with acceptable quality and within schedule based on the tender, stipulations, and site conditions.

5.4 IMPORTANT DECISIONS ARRIVED AT

Construction on both banks were to be mechanized and synchronized properly, and the main piers, P1 and P2, were to be started simultaneously along with A1 and A2 abutments.

To achieve the above progress and target as per the CA, all the necessary equipment, tools, plants, machinery, and vehicles were to be deployed and provided at the site within 30 days. These are identified and listed below.

(a)	compressor (300 cfm width), jack hammers, and all accessories	4 pcs
(b)	Poclain	2 pcs
(c)	drilling accessories	12 sets
(d)	vehicles for site movement	2 pcs
(e)	water pump 10 HP	6 pcs
(f)	hydraulic Excavators (JCB)	2 pcs
(g)	Tata tippers, 8 m^3 capacity	4 pcs

(h)	gen set, 15 kVA	2 sets
	125 kVA	2 sets
(j)	concrete mini batch plant	1 set
(k)	welding set	8 sets

All other vehicles/plants/equipment needed after completion of excavation are also identified. Hence, it was decided to place the order and ensure that items should be placed at the site on or before 30 August 2006.

concrete batch mix plant, 30 cum/ hr capacity	-	2 sets
concrete pump, 30 m³ /hr	-	2 sets
gen set, 125 kVA	-	2 sets
mobile crane, 20 t capacity	-	2 sets
tractor-cum-loader	-	2 sets
power press for cutting of steel bar	-	2 sets

A site laboratory was set up so that all required tests for the work could be carried out simultaneously to ensure proper quality by 30 June 2006.

A proper labour camp and an office were to be established by 30 June 2006. However, with the existing accommodation, the site office started functioning immediately after the acceptance of the CA.

The total required quantities of major construction materials like cement, steel, and aggregates were to be identified and orders to the concerned manufacturers had to be placed in a phased manner to avoid any slippage of delivery dates. It was ensured by the contractor that in every 15 days, the status of materials would be reviewed and that advance or corrective action would be taken up. The following are the quantities of the materials/work required:

Abstract of Quantity of Major Works

S/No	Items	A/U	Qty Assessed
1	excavation in foundation	cum	22150
2	lean concrete (M-15)	cum	124
3	concrete in foundation and substructure (M35)	cum	3150
4	concrete in pier and pier cap (M40)	cum	740
5	concrete in pier head and segmental construction (M50)	cum	4200
6	steel roads of various diameters	Mt	950
7	cement, OPC 43 Grade	Mt	4000
8	HTS wire, 15 mm GRPS	Mt	215
9	sheathing, 905 mm dia 0.5 mm thick	metre	1400
10	anchor bearing plate, including anchorages and cones	pcs	500
11	anchor cones	pcs	500
12	wedges	pcs	9000
13	bearing for piers and abutment, METCO Group	pcs	19 Set
14	STU Zerret of France supplied by METCO	pcs	4

The required quantity of shuttering materials had to be brought to the site before 30 June 2006, and shuttering were to be fabricated at the site only except CCSE, which was outsourced.

It was also decided that the order for bearings would be placed before November 2006, with the delivery by February 2007. And the order for cantilever construction shuttering equipment (two sets) was also to be placed, with the date of delivery at the site before 31 March 2007.

The payment of running bills was also raised by the contractor. The author, then commander, assured him of the disbarring of the payment within 24 hours on receipt of the correct running bill as per CA conditions. Also the TF had assured that the approval of the founding level will be given immediately on reaching designed RL.

The bridge had to be treated as two bridges on each bank that were to be later linked. Hence, all equipment, plants, vehicles, cops gantries, etc. had to be doubled and procured accordingly.

The work of the execution of the foundation (P1 and P2) was physically started on the ground by 5 June and 15 June 2006 respectively, and the plants and equipment were placed as per the original schedule. The excavation was only possible by drilling and blasting as very hard conglomerate strata in both the foundations (i.e. P1 and P2) was encountered after the removal of the initial overburden of earth up to a depth of about 3 m.

All the four foundations' excavations were carried out by continuous drilling and blasting due to the strata being extremely hard conglomerate soil. The blasting during excavation of the foundation was the only solution, and the same was only allowed with vigilance because the site is extremely near to the army and civil population.

The excavations for P1 and P4 and the completed plate load test to ascertain bearing capacity were carried out and are shown in the pictures below:

Fig. 20: View of excavation in foundation work.

Fig. 21: Completed excavation for lean concrete.

The geologist also contacted and organized a site visit. The geologist visited the site and submitted report and advised on how to proceed with the laying of concrete.

Fig. 22: View of load testing on completed foundation.

Fig. 23: Reading the results of the load test.

The lean concrete was laid after load testing, and the desired bearing capacity of the soil was achieved on 01 September 2006 at RL 311 and at RL 309.5 of piers P1 and P2 respectively.

The work was taken up in full swing for piers P1 and P2 and came to cross HFL. The reinforcement was placed in both piers' foundations, but heavy rains and high floods caused a setback because of floods and the pits being filled with mud, debris, etc. However, the surface was cleaned by deploying adequate people round the clock, and it was cleaned properly.

Fig. 24: View of reinforced laying of foundation.

Fig. 25: View of foundation ready for concreting.

The concreting of the foundations of pier P1 and P2 progressed and was completed by 14 October 2006. Work on both piers was continued with good speed. The P1 and P2 crossed HFL (at RL 330.4 m) on 11 December 06 and 14 December 06 respectively, and further progress was halted for want of bearings whose anchor bars were to have been embedded in the pier cap.

Fig. 26: View of foundation work in progress.

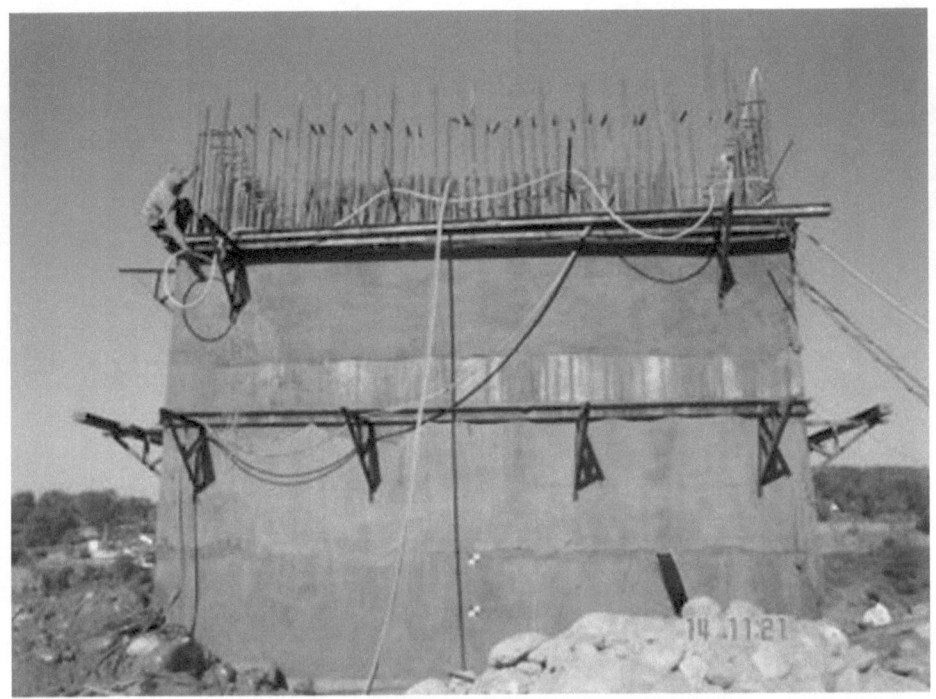

Fig. 27: View of pier work in progress.

The bearings were of a special type and extremely heavy, causing delay in manufacture vis-à-vis the time committed for delivery. The author had himself contacted the managing director of METCO, the manufacturer for the supply of the bearings, and came to know that they had some problems in machining the huge area of 1 m × 1.1 m, resulting in delay of three months, which delayed the assessed time for completing the bridge from 20 December 2007 to beyond March 2007. However, to avoid compounding delays and idling of resources, the work on the abutment of both sides was taken in full swing. The progress on the abutment A1 and A2 are shown here for clearer understanding.

Fig. 28: View of reinforcement work in progress.

Fig. 29: View of pier work in completion stage.

The first consignment of bearings reached the site on 5 March 2007, and all three sets of bearings were laid over P1 on 20 Mar 2007, and the base of the bearings on pier P1 was completed on 25 April 2007.

Fig. 30: View of laid bearing on pier head.

Fig. 31: View of finished surface after laid bearing.

The below the base of the bearing, concreting was carried out with self-compacted concrete. In this concrete, there is no need for the use of a vibrator. The method of laying and the final result are shown below:

Fig. 32: View of finished surface after pot bearing laid bearing.

The pier head construction was a difficult task. The erecting of the shuttering was started for casting of pier head P1 after laid bearing. The pier head was casted in three stages due its great height—up to a three-storey building. The special type of shuttering procured and modified at the site and finally casted was completed P1 pier head on 6 May 2007.

Fig. 33: View of bearing laid for fixing.

Fig. 34: Work of deck slab in progress.

The details and type of bearing procured and placed at the structure is tabulated below:

DETAILS OF BEARINGS

S/No.	Type of Bearings	Location	Qty (Nos)	Total wt in (MT)
01	Guide bearings	Centre of P2	02	10.60
02	Fixed pin bearings	Middle of P1	01	4.57
03	Elastomeric stopper bearing	Sides of A1	02	1.46
04	Elastomeric stopper bearing	Sides of A2	02	1.45
05	Pot bearings	Sides of P1	02	11.50
06	Pot bearings	Sides of P2	02	12.60
07	Biaxial pot bearing	Bottom of A1	02	0.80
08	Biaxial pot bearing	Bottom of A2	02	0.80
09	Biaxial pot bearing	Top of A1	02	0.50
10	Biaxial pot bearing	Top of A2	02	0.50
11	Shock Transmission Unit (STU)	Centre Reverse side of P2	04	2.42

The time cycle for the construction of the deck was optimized as six days per pair of segments. This was achieved by designing the concrete mix for the required strength at 48 hours to enable the first stage of prestressing.

The pier head construction for P2 was also simultaneously started, and all efforts were done to complete it immediately on receipt of the second and final consignment of bearing. The pier head for P2 was constructed on 9 June 2007.

The pier head for P1 and P2 were completed on 6 May and 7 June respectively. Thereafter, the segmental construction of the superstructure was continued, staging the shuttering on both piers by supporting from the ground because it was observed that the delivery of CCSE (composite construction of shuttering equipment) might be delayed, which may further delay the completion of the bridge. The casting for the segmental construction of the superstructure up to the fifth segment was completed by this process for both piers. Therefore, it can be said that 24 m of pier P1 and 24 on at pier P2; a total of 48 m in length of the superstructure out of the 280 m length had been completed. The fifth segment at piers P1 and P2 were completed on 01 July 2007 and 15 August 2007 and the load testing as well as stressing on 04 July and 18 August respectively. The results of the cube test after three days was 48 MPa and 48.8 MPa against 67% of M50 concrete. This result was achieved with OPC 43 Grade cement with 400 kg per m^3. The result had spired the ground executive and work had further decided to progress based on the three-day cube strength.

As soon as the construction of superstructure progressed, becoming reducing gap from the old standing pier, it was calculated that the old pier might have obstructed the casting of the superstructure; therefore, prior to reaching segmental construction nearer to the old pier, a minimum of 1.5 m from top of the old pier had to be cut. Since it was a difficult task, it was therefore decided to cut it with the help of an expert.

To have proper quality control and achieve good strength, the proper mix design was carried out at the site with the material being used. The results of the cubes test are shown below with the quantity of the materials used.

CUBE TEST RESULTS

Grade of concrete	Location where use	Design strength KN/m^2	Achieved after 28 days in KN/m^2	%age Over Achieved	%age use of Admixture
M15	Lean concrete in foundation	15	24.46	65	Nil
M35	Foundation & substructure	35	46.40	19	0.6% by Wt of cement Fosroc SP-430
M40	Pier & Pier cap	40	47.30	14	1.0% by Wt of cement Fosroc SP-430
M50	Pier head segment & Super structure	50	58.00	18	1.0% by Wt of cement Fosroc SP-500

Fig. 35: Results of the cube test.

CUBE TEST RESULT

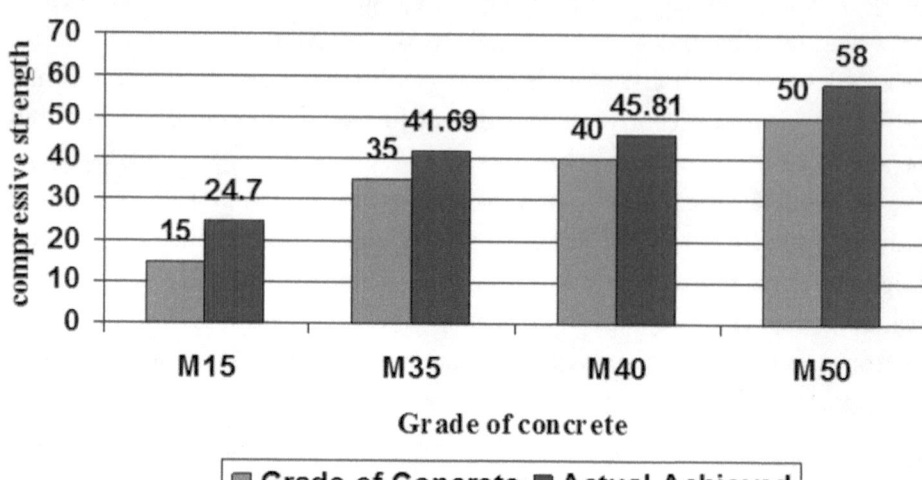

Fig. 36: Graphical view of the test results.

MIX DESIGN : TARGET VS ACHIEVEMENT

(Trial mix No 08)

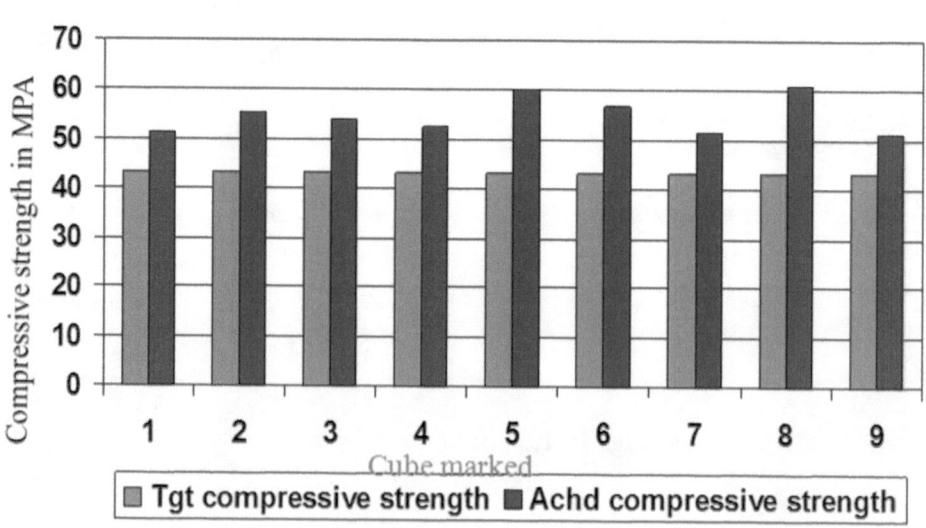

Fig. 37: Results of the graphical mix design.

CHAPTER

6

6.1 FOULING OF OLD STRUCTURES WITH THE NEW BRIDGE SCHEME

The proposed scheme for a new bridge consisted of the construction of a 280 m span balanced cantilever bridge. The bridge superstructure is balanced on the piers, cantilevering 80 m over the main stream and 60 m off the stream side to counterbalance on the abutment. With such a long span and obviously as per the technical requirement of the design and the deck level of the bridge fixed due to its approaches already constructed on acquired land, the depth of the superstructure was 10 m at pier location and gradually reduced as it spanned towards the centre of the bridge. With both the existing piers being at 11 m from the newly constructed piers, the soffit of the proposed superstructure was fouling with the existing piers. With no alternatives and with the constraints stated above, it was inevitable to dismantle the existing piers. The existing piers were to be dismantled to a depth of 1.5 m from the existing top level of the pier head.

6.2 POSSIBILITIES OF DISMANTLING

The first and foremost thought of a civil engineer is to go in for the breaking of the RCC structure by deploying manpower. This option was studied. The piers constructed were solid structures built of M35 concrete, 7 m long and 1.5 m wide, and with elliptical noses at the upstream and downstream end. Breaking the structures by hammering by manual labour was considered unsafe for men to work at an elevated height of 20 m above the ground near the river stream on skeletal scaffolding.

The second option was to break it with the help of rock breakers. This was also not considered effective due to the following reasons: heavy re-erection of scaffolding and the required working platform, congested reinforcing bars with large diameters induced in the structure, safety concerns for the men regarding the effect of vibration while working at an elevated height of 20 m above the bank of the river.

The third option was to drill holes and blast the old structures. This was straight away rejected due to the following reasons:

(a) The bridge site was located in the vicinity (about 10 m) of the archaeological exploration site Ambaran, and the archaeological department of the government of India had already issued a notice to avoid any blasting operation.

(b) The area was densely populated, with the civil populace in the far bank's end at Akhnoor and the army cantonment area on the home bank area. The flying concrete lumps due to blasting would also cause severe injuries and proved hazardous.

(c) The environmental factors.

(d) The difficulty in drilling holes with drills and in a structure heavily reinforced with steel and again taking into consideration the safety of the men working at an elevated height of 20 m from the ground.

6.3 PROPOSAL FOR DISMANTLING FINALIZED

With the said three options above not accepted, the willing of the unwanted portion of the piers was the only left out. Therefore, the search for firms that had the expertise to take up the task and come up with the best solution for the situation was started. Finally, after detailed enquires, the following firms were consulted:

(a) M/s Hilti
(b) M/s Binyas.

The site was inspected by the consultants, and a detailed survey was done to take up the job. The work was finally assigned to M/s Binyas, a Bangalore-based company. A detailed planning was done, and the consultants demonstrated a demo at the site, regarding the work to be done. They worked out the time involved and planned to complete the task in three consecutive days. The necessary equipment, machinery, tools, and plants were transported from Bangalore. The team of the firm consisted of an engineer and three technicians.

6.4 MACHINERY

The following machinery was brought by the consultant to the site:

(a) core drilling frame
(b) diamond wire saw machine for cutting and drilling (2,000 bar capacity, hydraulic, electrically operated with remote control)
(c) cylindrical core cutter with diamond teeth
(d) power pack make hydraulic pump with two hydraulic jacks capable of 60 mm lift.

6.5 PLANNING AND STAGES OF OPERATION

The ground level towards the new pier from the existing piers' location was RL 324.9 m, and the proposed cutting level of the pier was 333.4 m. Therefore, a working platform was required to be erected at a height of 7 m from the ground adjacent to the pier on the side off the stream as no space was available on the main stream's side and work could not be done from that end. The platform for working was erected using available steel cribs and steel shuttering plates.

The pier was proposed to be sliced off at about 1.5 m from the top by means of the diamond rope cutting machine, which was feasible with the machine. The cutting of the pier at the said level was to be done by the machine by revolving the rope around the periphery of the pier. The rope was driven by hydraulic mechanism. With no accessibility on the other side of the pier (as explained above, due to no space available on the main stream's side) and also with the machine incapable of cutting the entire 7 m length of the pier along the periphery in one go, further idea was developed.

6.6   DRILLING OF A HOLE IN THE CENTRE OF THE PIER

The cutting frame for a bore hole was fixed on the vertical face of the pier, with the four supports made in the pier. Then a 7 cm hole was drilled at the centre of the vertical face of the pier at the level proposed to be cut with a cylindrical diamond core cutter clamped to the frame (refer to the photo). As the cut top portion of the pier could also not be pushed down from the working platform due to friction and a total weight of 60 Mt, as planned earlier two holes were therefore drilled on the vertical face at the first/third and second/third on equal distances at the same level of the proposed cutting. The length of the holes was 50 cm deep inside the pier. All these operations took a total time of about one and a half hour with a cylindrical diamond core cutter.

6.7   CUTTING OF THE TOP PORTION OF THE PIER IN TWO PHASES

Thereafter, the bore hole's cutting frame was fixed on the face of the pier at the middle of the left portion of the centre hole drilled. The diamond core cutting machine was placed on the platform with an electric power supply with three phases. The diamond core cutting frame was fixed on the pier face below the level proposed to be cut. The special diamond-studded steel-wire rope just resembling that of a transparent round wire cable was passed through the central 7 cm dia hole drilled on the pier and brought around the pier to the other side so as to cover the periphery of the pier; it was positioned at the proposed level of cutting, fitted on to the core drilling frame, and tightened. After the setup, the electrically operated hydraulic machine was started, and the diamond steel rope started to revolve around the periphery of the pier in good speed, thereby starting to cut the RCC pier and embed 32 mm dia reinforcements.

6.8   PRECAUTION

The technicians and engineers were operating the machine through remote control and monitoring the operation from a considerable distance as a precaution to avoid any untoward incident in case of breakage of the diamond steel rope while in operation, which could cause serious injuries and even casualties. All spectators were not allowed near the machine as a precaution, as stated above. The entire operation lasted for one and a half hour. Similarly, the equipment was shifted to the right side of pier, and the other half of the pier was cut.

6.9 LIFTING OF THE CUT TOP PORTION OF PIER AND DISPLACING

After completion of the cutting operation, the machinery was removed, and two hydraulically operated jacks with capacity of 60 mm lift were placed on either side of the holes 18 cm and 50 cm in diameter drilled on the face of the pier. The jacks were connected by hose pipes to the hydrostress machine. Hydraulic pressure was to given to the jacks, which lifted the cut top portion of the pier, thereby trans-shipping its centre of gravity and gradually distancing the contact surfaces. The cut portion got misbalanced and fell down on the river bed. Similarly, the far bank's pier head was cut and dismantled.

Fig. 38: View of cut old pier being lifted.

The dismantling of the RCC pier head was carried out by Binyas as per the orders of the prime contractor.

6.10 CONCLUSION

The dismantling of the RCC pier head as illustrated above had been done in an manner that is environment-friendly. The cost of cutting such dense reinforced concrete structures are at the rate of Rs 20,000 per m^2. The total cost of the cutting work done was approximately Rs 4.2 lakhs, including overhead charges and transportation of the machinery to the site.

Even though the cost of dismantling on volumetric basis was Rs 8,750 per cum, which was considered to be high, it was the best-suited option according to the location and site of the work. There was no environmental disturbance or social problem and no effect on account of vibration, flying concrete pieces due to blasting, etc. on the newly constructed piers and abutment just adjacent to the dismantled one, and it was very safe. It so appeared that on completion of the work at the bridge site, a large white radish had been sliced from the top and had dropped down.

It was learnt that the adoption of such techniques would certainly turn out to be a revolution in the field of civil engineering and in the construction industry. The technique adopted was a non-destructive dismantling of structures.

CHAPTER 7

7.1 CONSTRUCTION OF THE SUPERSTRUCTURE (SEGMENTAL CONSTRUCTION)

At the conclusion of the cutting of the pier head of both piers and the completion of the fifth segment of the superstructure of both piers and the arrival of the CCSE at the site, immediate arrangements were made to push the work further.

The CCSE arrived at the site from the manufacturer without load testing. When the manufacturer was asked for load testing up to 130 Mt capacity, they then replied that there is no need as the equipment was manufactured to carry loads of more than 130 Mt. To avoid probable complications and have load testing for CCSE, the path was therefore subjected to load testing.

The undermentioned procedure was adopted as advised by the author at the site:

- CCSE was erected over the completed five segments of the superstructure of the bridge on both piers.
- The empty box of shuttering full suspended with CCSE was filled with sand up to a load of 130 Mt.
- The deflection of CCSE was measured and found to be 19 mm on full load.
- No fault was noticed during load testing at any joints of the CCSE.
- During concreting, the deflection of CCSE was observed around 10 mm after each segment casting.

The method adopted for load testing is shown below:

Fig. 39: Testing of CCSE.

Fig. 40: Measurement of deflection.

The segmental construction beyond the fifth segments of P1 and P2 were started upon receipt of the CCSE on 18 July and 15 August 2007 respectively. Efforts were made to speed up the work. Therefore, it was decided to have three stages of concrete cube tests (i.e. after 3 days, 7 days, and 28 days) to have early prestressing.

It was also decided that three-day and seven-day concrete cube should be kept on casted structure and its curing to be continued with the structure to ascertain the actual strength of the structure. It was observed that the three-day result was more than 67% (i.e. catering to the seven-day strength). Hence, it was decided that the stressing would be carried out after three days.

To have three-day strength, attempts were made to complete the casting of each set of segment on both piers within six days after getting 67% strength on three days. The plan accelerated the progress of the work and 75 days of delay were made up for.

The various activities involved and the actions taken to complete the construction cycle every six days are shown in the flow diagram.

HOW 06 DAYS CYCLE MOVE

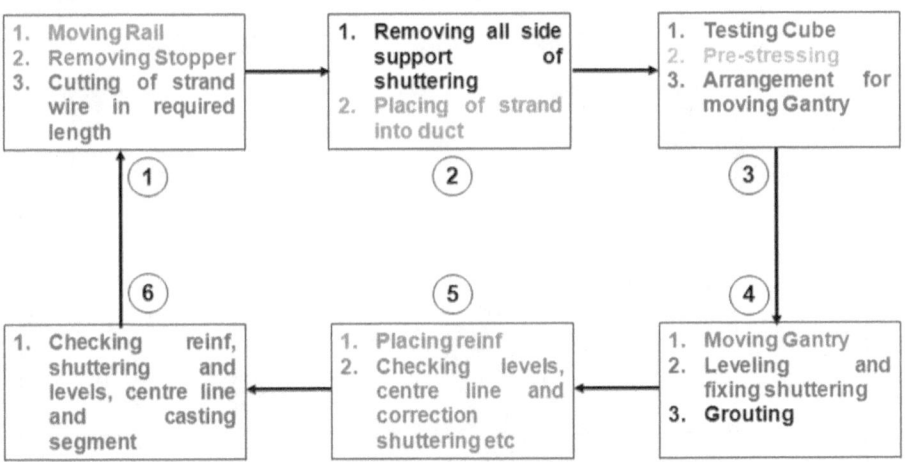

Note: - Independent activity for cutting, bending, binding, reinforcement, repair and cleaning shuttering and other processing works continue.

Fig. 41: View of six-day rotation of work for segments casting.

The details of the segments involved for construction, length, concrete quantities, stressing sequence, and cables are tabulated below:

SEGMENTAL DETAILS

Seg No.	Length of Seg (m)		Concrete Qty in Cum		Stressing sequence		Cable extension				
	Span 1 A/S	Span 2 R/S	Span 1 A/S	Span 2 R/S	Span 1 A/S	Span 2 R/S	Cable No.	Length of cable (mm)	As per design in mm	Phy Achd in mm P-1	P-2
1	2.50	2.50	104.00	104.00	-	-	-	-	-	-	-
2	2.25	2.25	50.52	50.52	31	31	1	10.70	76	78	80
3	2.25	2.25	48.16	48.16	1,2	1,2	2	15.01	108	111	111
										110	111
4	2.25	2.25	46.00	46.00	5	5	3	42.61	294	145	142
5	2.75	2.75	53.00	53.00	8	8	4	94.31	622	174	177
6	2.75	2.75	49.50	49.50	10	10	5	19.55	140	215.5	224
7	3.00	3.00	51.20	51.20	25	25	6	46.56	332	259	258
8	3.00	3.00	49.30	49.30	3	3	7	49.35	346	299	285
9	3.00	3.00	47.50	47.50	6	6	8	24.85	174	331	338
10	3.50	3.50	53.25	53.25	11	11	9	112.34	380	378	394
11	3.50	3.50	51.00	51.00	26,7	26,7	10	30.60	422	417	423
							11	55.64	346	342	340
12	3.50	3.50	49.20	49.20	12,21	12,21	12	69.53	464	474	471
							13	85.66	412	407	416
13	4.00	4.00	53.70	53.60	22,27	22,27	14	140.32	418	499	408
							15	131.37	516	404	519

Seg No.	Length of Seg (m)		Concrete Qty in Cum		Stressing sequence		Cable extension				
	Span 1 A/S	Span 2 R/S	Span 1 A/S	Span 2 R/S	Span 1 A/S	Span 2 R/S	Cable No.	Length of cable (mm)	As per design in mm	Actual in mm P-1	P-2
14	4.00	4.00	50.80	50.80	13,28	13,28	16	112.51	560	563	559
							17	112.57	558	570	554
15	4.50	4.50	55.10	53.90	4	4	E-1	59.90	622	617	598
							E-2	90.44	592		
16	4.50	4.50	53.30	51.10	29	29	E-3	82.94	662	689	648
					206		21	58.90	412		
17	4.50	4.50	51.20	48.70	9	9	22	59.90	714	715	698
					201, 202		23	96.42	630		
18	2.15	4.50	24.85	46.70	30,17	30,17	24	112.94	746	735	708
					204, 207		25	36.71	706	690	696
19	2.10	4.50	20.00	45.00	23	23	26	69.29	630	605	615
					203, 205		27	77.74	516		
20	-	4.50	-	43.60	15	15	28	85.62	830	799	
							29	103.65	662		
21	-	4.50	-	43.00	24	24	30	119.35	718	685	
							31	10.49	76		

Seg No.	Length of Seg (m)		Concrete Qty in Cum		Stressing sequence		Cable extension				
	Span 1 A/S	Span 2 R/S	Span 1 A/S	Span 2 R/S	Span 1 A/S	Span 2 R/S	Cable No.	Length of cable (mm)	Extension in mm		
									As per design in mm	Actual in mm	
										P-1	P-2
22	-	4.50	-	42.40	14	14	101	13.34	878	845	
23	- / 1.75		-	16.50	-	101, 102, 103, 104	102	49.33	338		
							103	13.34	96		
							104	31.34	218		
							105	49.36	336		
							106	76.38	500		
							108	67.38	450		
							109	58.36	396		
							110	22.38	160		
							111	58.36	396		
							112	22.38	160		
							113	40.37	280		
							114	67.38	450		
							201	14.36	104		
							202	14.36	104		

Seg No.	Length of Seg (m)		Concrete Qty in Cum		Stressing sequence		Cable extension				
	Span 1 A/S	Span 2	Span 1	Span 2	Span 1	Span 2	Cable No.	Length of cable (mm)	Extension in mm		
									As per design in mm	Actual in mm	
										P-1	P-2
							203	23.44	168		
							204	27.43	374		
							205	23.44	168		
							206	17.87	128		
							207	21.43	374		
							E-4	84.40	550		
							E-5	84.40	550		

To avoid excessive deflection due to unbalanced load during the construction, a properly designed bent on both sides of piers P1 and P2 towards the side of abutments A1 and A2 had been erected; whereas, in between P1 and P2 side pier, the existing old pier was used for this purpose. The neoprene pad was placed at the point of load transfer. The construction of the bent and the old piers used as bent are shown below:

Fig. 42: View of bent to control deflection.

Fig. 43: View if both side bent.

The segmental construction beyond the fifth segment was completed with very fast speed. The superstructure of the bridge between P1 and P2 and A1 was completed on 28 October 2007 just after the construction of

19th segment; whereas the superstructure of the bridge between P2 and A2 was completed by 24 November 07. The 22nd segment of P1 and P2 were also completed on 9 December 2007 and 16 December 2007 respectively, and finally, the 23rd segment (key segment) was cast on 30 December 2007, which linked the bridge.

Fig. 44: View of last segment.

Fig. 45: View of last segment stuttering.

Tabulated below are the details of the construction of each segment and its average strength after three days of concrete where stressing on the cable had been carried out.

DETAILS OF CONSTRUCTION AND STRESSING OF SEGMENTS

Segment No.	P1 side			P2 Side		
	Date of concrete	Date of stressing	Cube testing result (Avg)	Date of concrete	Date of stressing	Cube testing result (Avg)
Segment No. 2	25-04-07	28-04-07	41.19 Mpa	19-06-07	22-06-07	41.00 Mpa
Segment No. 3	10-05-07	13-05-07	37.5 Mpa	20-07-07	23-07-07	43.00 Mpa
Segment No. 4	23-06-07	26-05-07	43.26 Mpa	31-07-07	03-08-07	41.48 Mpa
Segment No. 5	01-07-07	04-07-07	48.00 Mpa	15-08-07	18-08-07	48.80 Mpa
Segment No. 6	18-07-07	21-07-07	53.00 Mpa	21-08-07	24-08-07	41.00 Mpa
Segment No. 7	27-07-07	30-07-07	40.59 Mpa	27-08-07	30-08-07	43.55 Mpa
Segment No. 8	02-08-07	05-08-07	40.00 Mpa	02-09-07	05-09-07	42.11 Mpa
Segment No. 9	08-08-07	11-08-07	43.00 Mpa	08-09-07	11-09-07	39.93 Mpa
Segment No. 10	15-08-07	18-08-07	39.60 Mpa	14-09-07	18-09-07	39.90 Mpa
Segment No. 11	22-08-07	25-08-07	43.56 Mpa	21-09-07	24-09-07	39.01 Mpa
Segment No. 12	28-08-07	31-08-07	40.11 Mpa	28-09-07	01-10-07	36.00 Mpa
Segment No. 13	04-09-07	07-09-07	40.00 Mpa	04-10-07	07-10-07	38.11 Mpa
Segment No. 14	13-09-07	16-09-07	40.59 Mpa	11-10-07	14-10-07	36.70 Mpa
Segment No. 15	21-09-07	24-09-07	39.48 Mpa	18-10-07	21-10-07	37.60 Mpa
Segment No. 16	28-09-07	01-10-07	39.56 Mpa	26-10-07	29-10-07	40.00 Mpa
Segment No. 17	05-10-07	08-10-07	38.15 Mpa	01-11-07	05-11-07	39.11 Mpa
Segment No. 18	12-10-07	15-10-07	37.48 Mpa	08-11-07	12-11-07	42.44 Mpa
Segment No. 19	28-10-07 (A/S)	31-10-07	44.55 Mpa	24-11-07	27-11-07	41.11 Mpa
	14-11-07 (R/S)	17-11-07	40.44 Mpa	01-12-07	05-12-07	42.50 Mpa
Segment No. 20	23-11-07	27-11-07	37.77 Mpa	08-12-07	12-12-07	42.22 Mpa
Segment No. 21	30-11-07	03-12-07	36.77 Mpa	15-12-07		
Segment No. 22	09-12-07		40.18 Mpa			

DEFLECTION

Segment No	P1				P2			
	RL As per Design		RL as per Actual		RL As per Design		RL as per Actual	
	A/S	R/S	A/S	R/S	A/S	R/S	A/S	R/S
1	342.453	342.576	342.444	342.529	342.453	342.576	342.433	342.578
2	342.397	342.630	342.385	342.598	342.397	342.630	342.378	342.643
3	342.340	342.682	342.304	342.665	342.340	342.682	342.299	342.713
4	342.284	342.733	342.257	342.726	342.284	342.733	342.245	342.775
5	342.215	342.792	342.205	342.781	342.215	342.792	342.198	342.846
6	342.147	342.850	342.109	342.871	342.147	342.850	342.111	342.916
7	342.072	342.909	342.048	342.966	342.072	342.909	342.053	342.975
8	341.997	342.966	341.961	343.016	341.997	342.966	341.959	343.034
9	341.922	343.021	341.897	343.081	341.922	343.021	341.876	343.090
10	341.834	343.080	341.805	343.148	341.834	343.080	341.795	343.168
11	341.747	343.136	341.711	343.197	341.747	343.136	341.704	343.218
12	341.659	343.188	341.619	343.248	341.659	343.188	341.621	343.249
13	341.550	343.243	341.502	343.312	341.550	343.243	341.512	343.317
14	341.459	343.292	341.408	343.369	341.459	343.292	341.405	343.366
15	341.347	343.342	341.312	343.413	341.347	343.342	341.317	343.421
16	341.234	343.386	341.194	343.444	341.234	343.386	341.200	343.451
17	341.122	343.423	341.099	343.458	341.122	343.423	341.095	343.505
18	341.068	343.454	341.054	343.476	341.068	343.454	341.049	343.564
19	-	343.479	-	343.522	-	343.479	-	343.589
20	-	343.497	-	343.541	-	343.497	-	-
21	-	343.509	-	343.566	-	343.509	-	-

7.2 FUTURE SUPPORT

To have future support for the superstructure upon occurrence of the destressing of the cables, the number 5 cable duct was left over.

ARRANGEMENT FOR FUTURE / EXTERNAL STRESSING

A. CABLE AT TOP (FOR DUMMY)

E-1	Segment No. 8 A/S to Segment No. 13 R/S	- 59.90 mtr
E-2	Segment No. 13 A/S to Segment No. 16 R/S	- 90.44 mtr
E-3	Segment No. 11 A/S to Segment No. 16 R/S	- 82.95 mtr

B. CABLE AT BOTTOM (FOR DUMMY)

E-4	Segment No. 13 to Segment No. 13	- 84.40 mtr
E-5	Segment No. 13 to Segment No. 13	- 84.40 mtr

C. ARRANGEMENT FOR FUTURE EXTERNAL STRESSING

05 Nos Duct left in Top of pier head (below the deck slab both P1 & P2

To have proper identification, the stressed cable up to the designed stress, the arrangement of blisters, the method of construction, and stressing in progress are shown below for clearer understanding.

DETAILS OF BLISTERS

Segment Nos	Cable at centre	No of cable stressed		Total Balance cable	New Addition		Bal	Blisters			
		Top	Bottom		A1-P1	P1-P2		A1-P1		P1-P2	
								Top	Bottom	Top	Bottom
1	62	·	·	62	·	·	·	·	·	·	·
2	62	2	·	60	·	·	·	·	·	·	·
3	60	4	·	56	·	·	·	⊡⊡	·	⊡⊡	·
4	56	2	·	54	·	·	·	·	·	·	·
5	54	2	·	52	·	·	·	⊡⊡	·	·	·
6	52	2	·	50	·	·	·	·	·	·	·
7	50	2	·	48	·	·	·	⊡⊡	·	·	·
8	48	2	·	46	·	·	·	⊡⊡	·	·	·
9	46	2	·	44	·	·	·	⊡⊡	·	·	·
10	44	2	·	42	·	·	·	·	·	·	·
11	42	4	·	38	4	·	42	⊡⊡⊡⊡	⊡⊡⊡⊡	⊡⊡	·
12	42	4	·	38	2	·	40	·	⊡⊡	⊡⊡	·

Segment Nos	Cable at centre	No of cable stressed		Total Balance cable	New Addition		Bal	Blisters			
		Top	Bottom		A1-P1	P1-P2		A1-P1		P1-P2	
								Top	Bottom	Top	Bottom
13	40	4	·	36	6	4	46	⊡ ⊡	⊡⊡ ⊡⊡	⊡⊡ ⊡⊡	⊡⊡ ⊡⊡
14	46	4	·	42	4	2	48	·	⊡⊡ ⊡⊡	·	⊡ ⊡
15	48	2	·	46	4	4	54	⊡ ⊡	·	⊡ ⊡	⊡⊡ ⊡⊡
16	54	2	2	50	2	4	56	⊡ ⊡	⊡ ⊡	⊡⊡ ⊡⊡	⊡⊡ ⊡⊡
17	56	2	4	50	·	4	54	⊡ ⊡	⊡⊡ ⊡⊡	⊡ ⊡	⊡⊡ ⊡⊡
18	54	14	4	36	·	2	38	⊡⊡⊡⊡	⊡⊡ ⊡⊡	⊡⊡ ⊡⊡	⊡ ⊡
19	38	2	4	32	·	2	34	⊡⊡ ⊡⊡	⊡⊡ ⊡⊡	⊡ ⊡	⊡ ⊡
20	34	2	·	32	·	4	36	·	·	⊡ ⊡	⊡⊡ ⊡⊡
21	36	2	·	34	·	4	38	·	·	⊡ ⊡	⊡⊡ ⊡⊡
22	38	2+EC6*	·	30	·	·	30	·	·	⊡ ⊡	·
23	30	30	·	·	·	·	·	·	·	·	·
	Total				22	30					

* Note: - EC = Emergency cable

The view of blisters during construction (already constructed and stressing in progress) is shown below.

Fig. 46: Casting of blister.

Fig. 47: View of blisters.

Fig. 48: View of prestressing cables.

Fig. 49: View of prestressed cable.

Quality control was strictly adhered to, and cubes were kept on the structure to ensure curing was carried out on the structure and cube with par to ascertain the actual result achieved by the structure. The strength achieved after 3 days, 7 days, and 28 days have been tabulated here, which clearly indicates the quality of the work.

7.3 STABILITY CABLE

The bridge is unbalanced cantilever, which behaves as a continuous cantilever bridge as soon as the superstructure of the bridge is linked. Therefore, it is very essential to protect the superstructure from uplift at abutment during construction due to unbalanced construction. Therefore, during construction of abutments, 2 sets of 19 cables at each abutment were provided to prevent uplift at abutment and were used to anchor the superstructure to the abutment. While span 1 was being completed, the superstructure was made to rest on a bearing on the abutment in order to progress the construction on the central part of the 40 m of span 2 (160 m). Stability cables have been stressed to 60% of its value.

These cables were to be removed after the stressing of two continuity cables. Remaining continuity cables were to be stressed only after stressing of the two continuity cables. A beam was provided on top to prevent an upwards thrust due to live load during service.

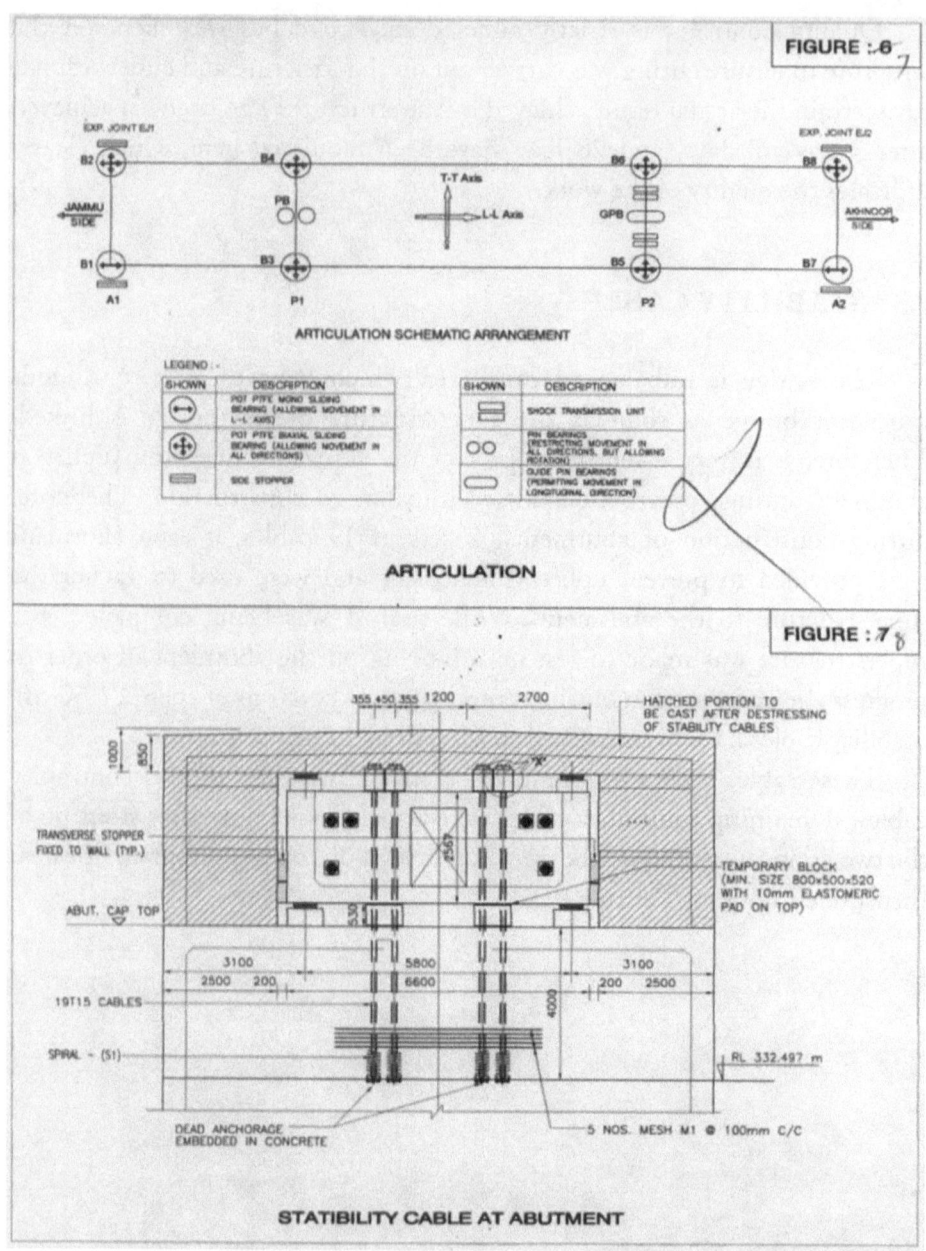

Fig. 50: View of stability cable.

The proper grouting was ensured after pre-cambering and stressing. The stressing sequence had been followed properly to avoid any obstruction during stressing. The proper homogenous mixing of grouting materials was ensured before being pumped into the duct of the cable. Stressing, mixing, and grouting of the cables are shown below for clearer understanding.

Fig. 51: Mixing of materials for grouting.

Fig. 52: Grouting of cables after prestressing.

The grouted duct was tested with the pressurized flow of water after grouting to ascertain possible leakage in the concrete.

The number 4 shock transmission unit were carefully fixed at P2 with the deck superstructure to absorb the excessive impact of STU before fixing and after fixing (shown below).

Fig. 53: View of transmission unit (STU).

The fixing of railings and crash barriers were commenced from both ends of the bridge (i.e. at A1 and A2) as soon as the superstructure rested on the abutment and stability cables were stressed up to the desired load. However, to save time, the railings were pre-casted at the site as per the approved drawing and were kept ready much before the completion of the segmental construction.

All efforts had been taken in time to avoid any delays in construction. The department had taken all precautions to ensure resources management at the site and RAR payments were outstanding. No decision was kept pending even for hours, and the topmost priority was given to the construction of this bridge. The payments submitted to the OC contract and the payments made by the commander for the contract are tabulated below:

DETAIL OF PAYMENT TO CONTRACTOR

S/No	RAR		RAR Amount (Rupees)				Amount paid (Rupees)	Date of Payment
	No	Date of submission	Work done (cumulative)	Advance for materials at site	Escalation (cumulative)	Total (cumulative)		
1	2	3	4	5	6	7	8	9
1	I	16.12.06	38978241	9544729	0	48522970	39260000	16.12.06
2	II	29.01.07	44429302	12111214	0	56540516	7489000	02.02.07
3	III	23.02.07	50890598	16413709	1959250	69263557	13397000	01.03.07
4	IV	27.03.07	62171557	22535790	3427869	88135216	16267000	28.03.07
5	V	26.04.07	65286449	24893245	3427869	93607563	4534000	01.05.07
6	VI	23.06.07	83133938	26163371	5509404	114806713	19841000	26.06.07
7	VII	23.07.07	90900121	28813115	5381229	125094465	8963000	24.07.07
8	VIII	24.08.07	116071814	23048014	5423379	144543207	19638000	27.08.07
9	IX	24.09.07	148357247	19338162	9101002	176796411	31085000	26.09.07
10	X	25.10.07	171276958	12185318	9059122	192521398	1649800	26.10.07
						Total amount paid	176972000	

The approval for the various stages of work were made without any time gap between readiness and decisions. The four levels of foundation were approved on the same day by the accepting officer. The stages are shown below.

STAGES APPROVED BY AUTHORITY				
S/No.	Date	Important activities in the stage	Approving authority	Recommendation by
01	28 April 2006	Inauguration; laying of foundation stone	Hon'ble Chief Minister Janab Ghulam Nabi Azad	Commander of TF, B. S. Pandey
02	17 August 2006	Plate load test on P1 and approval of fdn	Brig V Rajagopal CE (P) Sampark	Commander of TF, B. S. Pandey
03	30 August 2006	Approval of fdn P2	Brig V Rajagopal CE (P) Sampark	Commander of TF, B. S. Pandey
04	09 November 2006	Approval of fdn P2	Brig V Rajagopal CE (P) Sampark	Commander of TF, B. S. Pandey
05	14 December 2006	Approval of fdn P1	Brig V Rajagopal CE (P) Sampark	Commander of TF, B. S. Pandey

8.1 ADMIXTURE

A few admixtures were used to increased workability. The details of the admixtures were already provided by the manufacturers. The quantities used and the effects observed are tabulated below:

ADMIXTURES

1. CONPLAST SP-430 -   1% by wt of cement
used in M40 concrete

ADVANTAGES

❖   High range water reducer
❖   Very high workability aid
❖   Provides high water reduction without loss of workability
❖   High ultimate strength can be achieved

2. CONPLAST SP-500 -   1% by wt of cement
used in M50 concrete

ADVANTAGES

❖   To produce pumpable concrete
❖   To produce high strength, high grade concrete by substantial reduction in water resulting in low permeability and high early strength
❖   To produce high workability concrete requiring little or no vibration during placing

3. CONBEXTRA GP2: -   **Free flow high strength concrete**
                                                   (Self compacting concrete)
                                                   Water content 18.00% of mix
Make   -   Fosroc

ADVANTAGES: -

❑ **Gaseous expansion system compensates for shrinkage and settlement in the plastic state**
❑ No metallic iron content to cause staining
❑ **Pre-packed materials overcomes onsite batching variations**
❑ Develops high early strength without the use of chlorides
❑ **High ultimate strength ensure the durability of the hardened grout**
❑ Free flow ensures high level of contact with load bearing area

4. CEBEX 100 -   **Admixture for grouting**

Make   -   Fosroc

Advantages: -

❑ **Gaseous expansion system compensates for plastic shrinkage and settlement in properly designed cementitious grout**
❑ Reduced water/cement ratio mixes in the grout mix ensures low permeability and long term durability in service
❑ **Gives high grout fluidity with low water/cement ratio, thus making placement or injection of the grout easy**
❑ No metallic iron content to corrode and cause staining or deterioration due to rust expansion in the grout
❑ **Composition allows high early strength development in grouts, without the use of chlorides**

5. CONCURE WB - **Concrete curing compound to restrict evaporation of water during high temperature**

Make - Fosroc

<u>Advantages: -</u>

❑ **Complies to ASTM bC309-98**
❑ Single application – spray applied to fresh concrete, forms a barrier to water loss
❑ **Eliminates water curing where difficult to apply**
❑ Reliable – no risk of erratic or poor curing
❑ **High quality surface – minimizes risk of drying shrinkage cracks and dusty surfaces**

6. CONPLAST SR - **Water-based surface retarder for concrete to provide good bond between two segments**

Make - Fosroc

<u>Advantages: -</u>

❑ **Economical in application**
❑ Simply applied, straight from the container
❑ **Need not use a mould release agent**
❑ Good results with heat curing and high temperature concreting
❑ **Can be used on steel, glass, fibre or timber formwork**
❑ **Non-toxic and non-flammable**

8.2 LABORATORY AT THE SITE

A proper laboratory was established at the site by the prime contractor for proper testing of all ongoing work to ensure proper quality of the work. The details of the equipment are shown in Appendix A.

8.3 VEH/PLANTS/EQUIPMENT

The vehicles, plants, and equipment deployed to complete this bridge are listed at Appendix B (with capacity).

8.4 MAJOR CONSTRUCTION STORE

Proper storage facilities were created before starting the work to avoid environmental effects. Special care was taken on the proper stacking of all materials. The material report on the stores are shown in Appendix C.

CHAPTER

9

9.1 INNOVATION

A number of innovations were achieved during the design and construction of this bridge.

The innovations/firsts are listed below:

* This is the longest-spanning prestressed concrete bridge (160 m) in India.
* This is the fastest construction of any PSC structure in India for a bridge 280 m length.
* This has the shortest time cycle consistently achieved for the construction of each pair of segments (six days).
* This has the heaviest bearings ever used in any bridge in India. Total weight of bearings (42.00 Mt), maximum capacity (4,000 Mt), physical weight of the bearings are as follows:

S/No.	Type of bearings	Location	Qty.	Total Wt in (MT)	Remarks
1.	Guide bearing	Centre of P2	2	10.6	Design capacity of the bearing is 4000 t
2.	Fixed pin bearings	Middle of P1	1	4.57	
3.	Electrometric stopper bearing	Sides of A1	2	1.46	

S/No.	Type of bearings	Location	Qty.	Total Wt in (MT)	Remarks
4.	Electrometric stopper bearing	Sides of A2	2	1.45	Design capacity of the bearing is 4,000 t
5.	Pot bearings	Sides of P1	2	11.5	
6.	Pot bearings	Sides of P2	2	12.6	
7.	Biaxial pot bearings	Bottom of A1	2	0.8	
8.	Biaxial pot bearings	Bottom of A2	2	0.8	
9.	Biaxial pot bearings	Top of A1	2	0.5	
10.	Biaxial pot bearings	Top of A2	2	0.5	
11.	Shock Transmission Unit (STU)	Centre Reserve side of P2	4	2.42	
				23	

- ❖ Concrete was mixed with minimum cement content of 400 kg (M50/M43 Grade) used in India, along with admixture (cement used was OPC 43 Grade).
- ❖ Fe500 steel for reinforced bars was used for the first time in India in a cantilever construction bridge.
- ❖ Prestressing was done after only 60 hours of casting of each segment.
- ❖ The concrete below the bearings were done using Conbextra grout for the first time. Self-compacting concrete (SCC) was used for the first time for concrete below the bearings.

Use of quick-flow cement.

- ❖ The depth of the girder in the central portion is 3.4 m for the 160 m span. This has a high span depth ratio of more than 47.
- ❖ Total concreting was done automatic-batching concrete plant and concrete pump. Concrete was mixed, transported, and placed untouched by hand for a cantilever construction river bridge in India.
- ❖ Electrical passenger hoist for inspection was used for the first time in India for this river bridge.
- ❖ Surface retarder and curing compounds were used for the cantilever bridge.
- ❖ For the first time, 43 Grade cement was used for 50 MPa concrete.
- ❖ This had the fastest approval for constructions of foundations.
- ❖ Segments on both sides of the pier were concreted simultaneously, balancing the weights by doing synchronized casting for the first time.

9.2 LESSON LEARNT

There are a number of lessons learnt from the construction of this bridge.

- ❖ On hindsight, the repeated changes of the GAD by changing the locations of the foundations should have been avoided. Proper analysis was required when the tilting of wells occurred repeatedly in the stream rather than putting more wells in the flowing zone of the river.

- ❖ As per the site conditions and ground strata, the best-suited foundation was open, which was taken care of when the new concept for cantilever construction was proposed.

- ❖ For the completion of the bridge, it was essential to complete the foundations first rather than the casting of superstructure elements. This was ensured in the new scheme.

- ❖ Construction stores and resources management should be critically examined, maintained, planned, and all precautions/remedial measures need to be made in time to avoid critical loss of time.

- ❖ To ensure the timely completion of the project, approval of the different stages of works by the competent authority were monitored, apart from the payment of running bills. Bills were paid within 24 hours of receipt. The progress was closely monitored at all formations. The period of completion was sacrosanct, as the defence minister had announced on the floor of the Lok Sabha.

- ❖ Heavy floods on 3 August 2006 and 2 September 2006 caused damages to the cofferdam, and the entire excavated foundation was filled with mud and debris twice, causing a delay in the completion of the foundation. Hence, it is essential to take proper care of such foundations from flash floods by having suitable cofferdams.

Fig. 54: Completed foundation flooded.

Fig. 55: Cleaning mud from the foundation.

❖ The concept of using more cement for achieving greater strength should be reviewed. Greater strength is possible by proper control of water–cement ratio and use of nominal and requisite quantity of admixture to increase the workability and use of proper grade of aggregates as per the Mixed Design.

9.3 LOAD TEST ON SUPERSTRUCTURE

Fig. 56: Load test on superstructure.

Fig. 57: Deflection measurement.

CHAPTER

10

10.1 VISIT OF VIPS AND THEIR REMARKS

A large number of VIPs visited the site, and their views/opinions are given below. The visit was organized properly and a demo of the work in progress was given apart from a briefing of the entire project with the completion costs. The list of VIPs and their comments are given below:

S No	Date	Name of VIP	Rank	Remarks
1.	28 Apr. 2006	Jenab Ghulam Nabi Azad	CM of J & K with ministers and MPs/MLAs	
2.	9 May 2006 onwards	B. S. Pandey	SE (Civ), CDR, 13 BRTF	Visited every day during construction
3.	16 June 2006, 20 Nov. 2006, 7 Feb. 2007, 10 June 2007, 29 Aug. 2007	Lt. Gen. K. S. Rao	PVSM, SC, SM, DGBR	'A well-planned and excellent site organization. Keep up the steady progress.'
4.	16 June 2006; 10, 17, 25, 30 Aug. 2006; 9 Nov. 2006; 14 Dec. 2006; 24 Mar. 2007; 21 May 2007; 2 June 2007; 24 June 2007; 24 Aug. 2007; 29 Aug. 2007; 31 Aug. 2007	Brig V Rajagopal	CE (P) SPK	

S No	Date	Name of VIP	Rank	Remarks
5.	14 Aug. 2006, 9 Nov. 2006, 24 Mar. 07, 2 June 2007, 14 July 2007, 22 Sept. 2007	Shri S. A. Reddi	chief consultant	'Excellent quality, shortest segmental time cycle, dedicated project team, including subcontractors in shaping a project of many fists in the country.'
6.	7 Feb. 2007, 21 Aug. 2007	Lt. Gen. Anil Malik	GOC, 10th Infantry Division	'Best of luck. Great job.'
7.	23 Feb. 2007	Shri Murlidhar Pandey	director, BRDB	
8.	11 Mar. 2007	Shri Alok Majumdar	senior installation engineer, METCO	
9.	25 Mar. 2007	Shri M. L. Sharma	MP (J & K)	
10.	25 Mar. 2007	Shri Shyam Lal Sharma	MLA (J & K)	
11.	18 May 2007, 29 Aug. 2007	Shri Pawan Kotwal, IAS	secretary to chief minister (J & K)	'I am quite impressed with the pace and quality of work of the bridge. I wish, with your dedication, it would be commissioned by the end of this year.'
12.	2 June 2007	Shri T. P. Velayudhan	ADG (BR)	
13.	2 June 2007	Dr V. K. Yadav	DDG (bridge)	
14.	2 June 2007	Shri Rajeev Ahuja	proof consultant, Arch Infrastructures Pvt Ltd	

S No	Date	Name of VIP	Rank	Remarks
15.	2 June 2007	Shri R. B. Singh	designer of bridge, DPDC	
16.	2 June 2007	Shri R. K. Gupta	proprietor of CCSE	
17.	2 June 2007, 14 July 2007	D. D. Sharma	chairman, D2S Infrastructures Pvt Ltd	
18.	10 June 2007	Shri Hari Charan Jeet Singh	secretary	
19.	21 July 2007	Col L. Padmanabhan	TS to DGBR	
20.	24 Aug. 2007	Mrs Soumya Rajagopal		'Amazing experience going up in the "hoist", and hats off to BRO and their personnel.'
21.	24 Aug. 2007	Mrs Mamta Varma		'Thrilling experience on the *hoist*. Seeing all these, I felt good.'
22.	29 Aug. 2007	Shri Pawan Kotwal, IAS	secretary to CM J & K	'I am quite impressed with the pace and quality of work on the bridge. I wish, with your dedication, it would be commissioned by the end of this year.'
12.	5 Sept. 2007	Maj. Gen. A. K. Manan	major general, CE Western Command	'An outstanding project. God bless BRO.'
24.	8 Sept. 2007	Brig V. K. Yadav with ten officers	SC, Artillery Div.	'Marvellous technologies. May God succeed.'
25.	10 Sept. 2007	Brig B. S. Chauhan, SM	HQ, 191 Infantry Brigade	'Great engineering feel. We are proud yet.'

S No	Date	Name of VIP	Rank	Remarks
26.	13 Sept. 2007	Mrs Geetha Rajagopalan	family member, CE	'Excellent.'
27.	13 Sept. 2007	Vasantha Sesjadu	family member, CE	'Marvellous. Wonderful.'
28.	13 Sept. 2007	V. Seshadr	family member, CE	'A challenging work being carried out authentically. All the best for a successful completion.'
29.	13 Sept. 2007	Rajagopalan, RV	family member, CE	'Hats off to Sampark and his team. Wonderful design and execution.'
30.	19 Sept. 2007	Lt Gen T. K. Sapru, YSM	GOC, 16 Corps	'It has been a great education to learn the process of making such an engineering marvel. Wish you all the best, and hope you do provide the link over the wonderful Chenab on time. God bless.'
31.	22 Sept 2007	Shri M. C. Boro	secretary, PWD of Assam	'It is the best engineering structure made by Indian engineers. It may be taken as an example for future bridge constructions.'
32.	22 Sept. 2007	Shri Sunil Kumar Sharma	deputy secretary, PWD of Assam	'I would like to thank Shri D. D. Sharma for giving me an opportunity to visit the RCC bridge construction over river Chenab along with my engineering

S No	Date	Name of VIP	Rank	Remarks
				bridge design officer. This is an engineering marvel in bridge engineering, and Mr Sharma has brought pride to the engineering fraternity of Assam PWD, with which he has a long association. My engineering colleagues are going to get a large exposure in the way of experience in bridge engineering.'
33.	22 Sept. 2007	Shri Syed Mainuddin Hassan	EE, PWD (R), CE Office, Guwahati	'I would thank Shri D. D. Sharma for arranging the visit to the bridge site, which is one of the most challenging engineering tasks. I am really overwhelmed by the minute design detailing being executed in the site. This visit will remain in my mind, and the experience I gather will be very beneficial for my career as a bridge designer. I thank D. D. Sharma again.'

S No	Date	Name of VIP	Rank	Remarks
34.	22 Sept. 2007	Shri D. Gogoi	EE, PWD, Guwahati	'An exciting and challenging task to construct a bridge over the river Chenab under the documented circumstances. Hats off to the BRO and M/s D2S Infrastructures for their excellent job.'
35.	22 Sept. 2007	Shri S. A. Reddi	chief consultant	'Excellent quality, shortest segmental time cycle, dedicated project team, including subcontractors in shaping a project of many fists in the country.'
36.	21 Oct. 2007	Shri A. K. Bharati	Addl CE PWD (RTO); presently, CE of Assam Police Housing Corp.	'Excellent work D. D. Sharma has done. It is an achievement that the work of the bridge is being completed within the target date of completion. The bridge itself involves high technology with segment construction with many more perfect technology applied in the construction. I wish him good luck for more achievements in the days to come.'

S No	Date	Name of VIP	Rank	Remarks
37.	6 Dec. 2007	Shri Sanjay Jain	director, Arch Consultancy Services (P) Ltd	'As a designer, it is a good experience to see these kinds of structures which we see only on drawings. The quality of the construction is excellent. Finishing too is good. The structure very well matches with the surroundings. Compliments to the contractor, and thanks for making this visit possible.'
38.	6 Dec. 2007	Shri Rajeev Ahuja	Managing director, Arch Consultancy Services (P) Ltd	'Speed of construction completed with good quality is really impressive.'
39.	8 Dec. 2007	Mohd Hussain	MOSH Earth and Power	'I am very happy to see the development in record time. The work is excellent and in record time. The agency needs facilitation.'
40.	8 Dec. 2007	Shri Ram Lal Sharma	executive magistrate, first class, Akhnoor	'This is best company I have ever seen. Its construction work and speed of work is perhaps unique in India. I have experience of seeing Rashtriya Chemicals & Fertilizers Limited Jhat, Raigarh in

S No	Date	Name of VIP	Rank	Remarks
				Bombay when I was in C/SF. This has influenced me very much.'
41.	9 Dec. 2007	Shri H. C. Arora	CE MOSRTRH, New Delhi	'I am impressed with the pace of work and quality of work. I wish that some provision of external prestressing should have been kept. Overall management and work is excellent.'
42.	9 Dec. 2007	Shri Dheeraj	executive engineer, MOSRTRH, New Delhi	'The work is in good progress, but I think the safety measures should be improved with safety shoes, belts, etc. Thanks for the experience shared by the chairman today.'
43.	9 Dec. 2007	Shri Abhijeet	assistant executive engineer, MOSRT&H, New Delhi	'The progress of work is to be noted, and the quality of work and methodology adopted to maintain the quality is highly appreciated. Thanks for sharing the experience and the difficulties faced during execution.'
44.	10 Dec. 2007	Shri Chamanlal	CE, MORTH	'Quality of work is very good, and the speed with which progress is achieved is wonderful. It would

S No	Date	Name of VIP	Rank	Remarks
				have been very fast if the end span were kept as half of the main span. I wish the team involved all success.'
45.	10 Dec. 2007	Shri D. K. Sharma	SE, MORT&H	'Bridge construction under leadership of Shri D. D. Sharma has been really organized and as per the ethics of engineering profession. It is reflected in achieving the completion of the work and that too is a style.'
46.	16 Dec. 2007	Dr R. C. Sharma	minister of state for Industries and Commerce (PHE, IFC)	'Today on 16 December 2007, I visited the Chenab bridge, Akhnoor. The works is well executed, and I am very happy to see the construction. As per the statement given by Mr Joginder Singh, they will complete the bridge by the end of December 2007. I congratulate D2S Infrastructures that they are going to hand it over in time.'

A briefing was organized at the site on the visit of the VIPs during finalization of the scheme and its construction. The discussions were captured at various stages and are shown below.

Picture 1: Briefing of DGBR and CE regarding new scheme.

Picture 2: Briefing of DGBR regarding visibility of new scheme.

Picture 3: Briefing of CM (J & K) regarding new view scheme.

Picture 4: Briefing of CM (J & K) regarding laying of scheme on the ground.

Picture 5: Discussion regarding difficulties faced
during execution of foundation.

Picture 6: Site visit by DGBR and GOC (10 Div) and designer of the bridge
with contractor, chief consultant, and discussion of the problems at the site.

Picture 7: Inspection of pier work by all dignitaries.

Picture 8: Inspection of work by CE, Cdr, and chief consultant.

Picture 9: Puja while construction of superstructure started.

Picture 10: Inspection of superstructure started
by local MLA and MP with Cdr.

Picture 11: Visit at the site by PS and CM and discussion about progress of work.

Picture 12: Inspection of work by ADGBR and chief constructor.

Picture 13: All dignitaries being briefed by Cdr regarding progress.

Picture 14: Visit of Sect BRDB and discussion regarding progress.

Picture 15: DGBR watching the progress of work on the movement of gantry for further deck slap top.

Picture 16: Gen. Manan observing segment curing.

Picture 17: Cdr briefing Lt. Gen. (16 Corps), GOC
(10 Div) regarding stage of construction.

Picture 18: Generals are observing the various stages of segmental construction.

Picture 19: Both generals are discussing the methodology of quality, and dedication and appreciating the progress, of BRO.

Picture 20: Both generals are walking on casted deck slab segmental construction.

Picture 21: Secretary of Assam's PWD with
contractor observing work at the site.

Picture 22: Secretary of Assam's PWD and others.

11.1 CONCLUSION AND RECOMMENDATIONS

Technical and administrative decisions should be taken on merit and based on the available technology after proper analysis. By taking bold decisions not based on hiding or camouflaging the acts of the past mistakes, this bridge has seen the light of day. Our felicitation to all concerned.

Sites where the velocity of the flow of water is high (i.e. more than 4 m/s) and where the bed is conglomerate, soft stone, bouldery, etc.—in short, difficult to penetrate—such types of constructions of continuous long-spanning bridges are recommended.

Other countries are achieving constructions with spans of around 200 m. We should also explore constructing such long spans.

The grade of the concrete designed for use in the superstructure of this bridge was M50. It is recommended that for such large spans, grades of around M80 and above should be targeted. This will reduce the cost of construction as well as the dead load of the superstructure. Incidentally, in this bridge, M50 was used in the first trial mix with minimum cement of 400 kg/m³ content and normal admixtures.

Continuous survey of the levels and centre line with the help of total station is a must for these types of structures. In this bridge, continuous checking of deflections and the centre line as per the approved drawings conceived by the design consultant and as actually observed were performed and these matched to a T. There were no variations. Thus, the linking segment was cast at the same level on both ends in elevation with a matching centre line.

Time is of the essence for the contract, and money is the cause of it. The department, by its actions in this bridge, has shown that in spite of stiff financial conditions, there were no conflicts, and perfect harmony was maintained during execution.

For important time-bound projects, a team of officers to start the project should be made to complete the same to ensure continuity of command, accountability, efficiency, and satisfaction of the team.

When the 22nd segments were facing each other and the shuttering of the 23rd segment (i.e. the linking segment) was to have been placed, there was no level difference, and the levels matched on both tips to the nearest millimetre both in plan (centre line) as well as in elevation. This was possible because every day the levels were maintained by a team of surveyors with the help of the total station. These levels were sent to the design consultant, who monitored these personally. In fact, after the concreting of each segment, the levels actually measured and envisaged by the designer fitted almost to a T. This proves that the parameters fixed by the design consultant and the parameters actually achieved during execution were complementing each other. The cables were so placed that almost all the cables were straight and without any kink. Thus, prestressing results were exactly as shown in the approved drawings both in terms of extension and gauge pressure.

The area where the bridge is located experiences extreme weather (extreme cold and extreme hot weather temperature vary from 5 °C to 46 °C). Concreting in this extreme temperatures can become risky as extreme changes in temperatures can result in microscopic cracking as the steel shuttering becomes very hot. However, it is in extreme cold temperatures (below 10 °C) when the cube results drastically fall from their targeted strengths. In this bridge, while concreting during the day, concreting was completed by 11 a.m., starting it early in the morning, or if started in the evening, it is completed in the night. During cold weather, the water was heated with electrical heating immersion rods or heating the same in preheated water tanks by burning firewood below dams. By this action, the cube results were achieved.

On the advice of the proof consultants, the reinforcements as required in the detailed design in this bridge was increased substantially in the blisters. This led to early stressing without any problems/de-stress. It is recommended that reinforcements in blisters should be increased.

The bridge scheme was unequal, the river side's span (half) was 80 m and the land side's was 60 m. Thus unbalanced moments had to be accommodated in the land side by additional dead load. The difficulty in construction and the risk of unequal moments could have been avoided if the land side's span was made equal (i.e. increased from 60 m to 80 m). This action would have reduced the risks without any increase in costs as executed structure had casted more.

The designer of this bridge had scrupulously followed the use of tapering sections, haunches, and chamfers for better and smoother flow of forces to avoid stress concentration and/or cracking. This method, although required as per code, has almost been forgotten. It is recommended that wherever force transfer is taking place, the old, tried, and tested methods of easing of stress should be scrupulously followed.

During the last decade, technology has advanced exponentially. Hence, for all problematic bridges lying incomplete elsewhere, the adoption of the latest technology will definitely expedite completion as has been achieved in Chenab Bridge.

APPENDIXES

APPENDIX A

LABORATORY EQUIPMENT

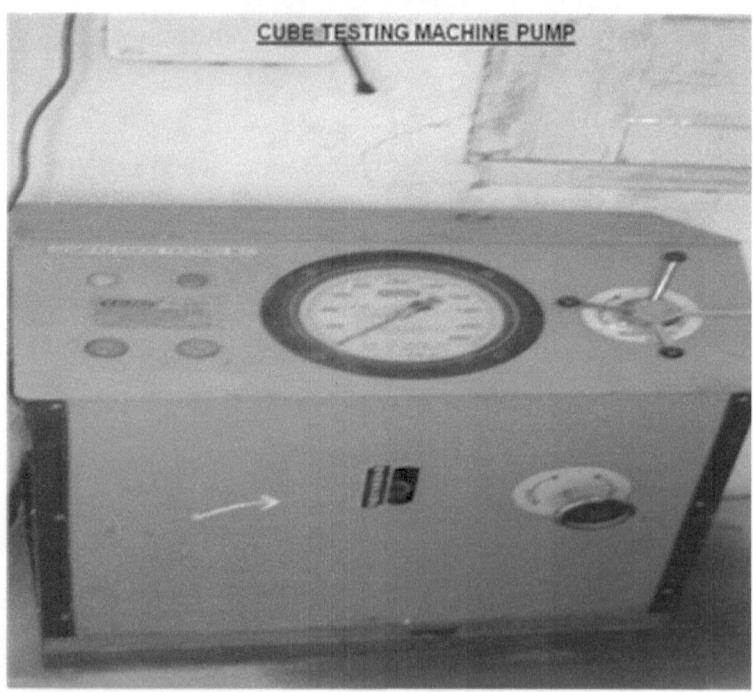

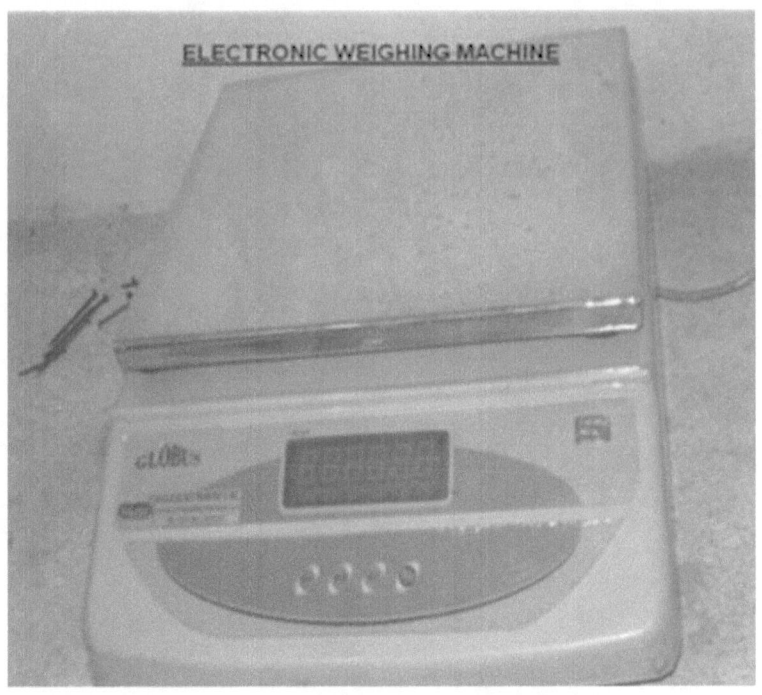

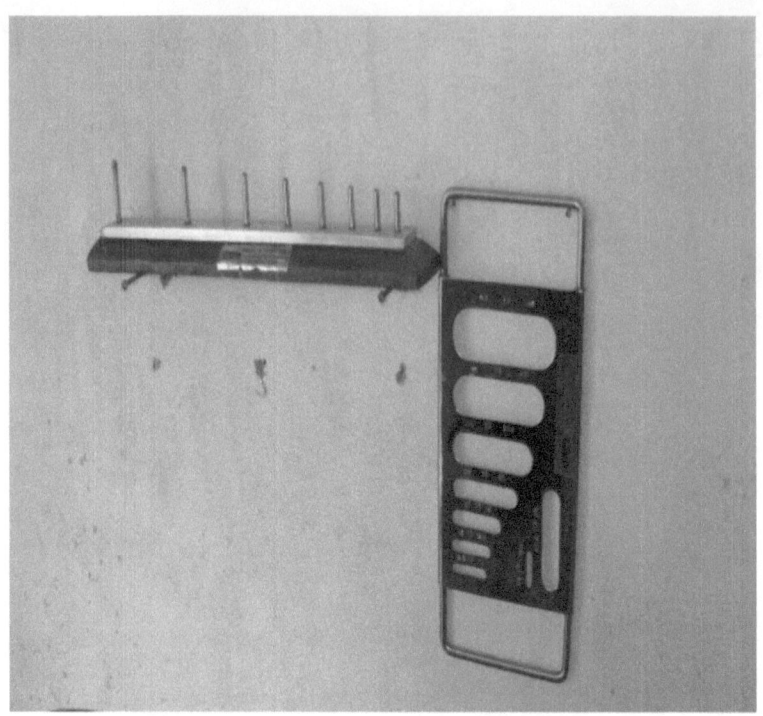

APPENDIX B

VEHICLES/PLANTS/EQUIPMENT

HYDRAULIC EXCAVATOR JCB

MAKE : JCB   CAP : BUCKET SIZE 0.75 CUM

HYDRAULIC EXCAVATOR (POCKLAIN)

MAKE : L&T

APPENDIX C

MAJOR CONSTRUCTION STORAGE

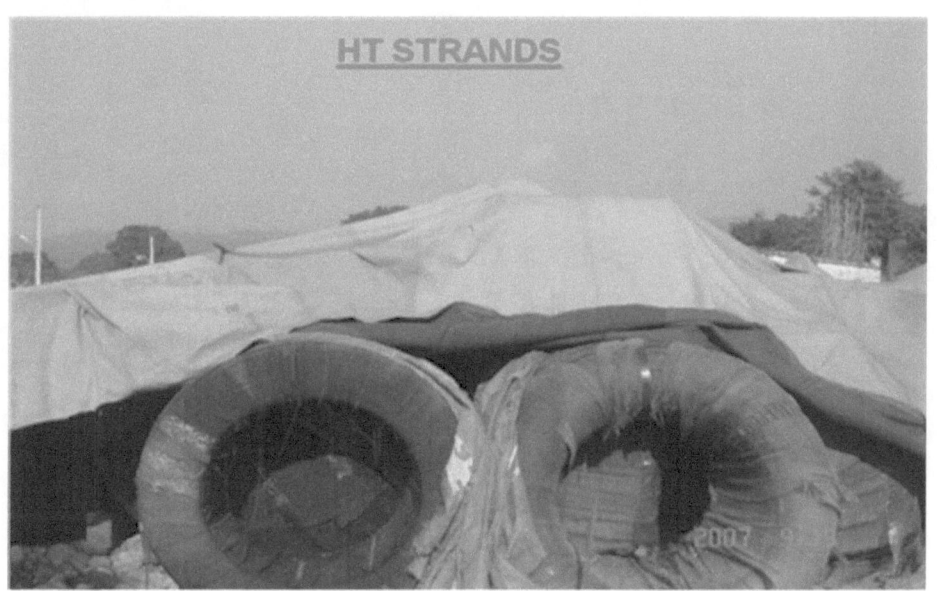

www.ingramcontent.com/pod-product-compliance
Lightning Source LLC
Chambersburg PA
CBHW021952170526
45157CB00003B/953